Lecture Notes in Economics and Mathematical Systems

553

Burkart Mönch

Strategic Trading in Illiquid Markets

 Springer

Author

Burkart Mönch
Goethe University Frankfurt
School of Economics and Business Administration
Finance Department, Derivatives Group
Uni-PF 77
60054 Frankfurt am Main
Germany
E-mail: b.moench@econ.uni-frankfurt.de

Library of Congress Control Number: 2005922554

ISSN 0075-8442
ISBN 3-540-25039-5 Springer Berlin Heidelberg New York

Springer is a part of Springer Science+Business Media

springeronline.com

© Springer-Verlag Berlin Heidelberg 2005
Printed in Germany

Typesetting: Camera ready by author
Cover design: *Erich Kirchner*, Heidelberg

Printed on acid-free paper   42/3130Di   5 4 3 2 1 0

Acknowledgments

This work arose from three papers that have been written while I have been a research assistant at Johann Wolfgang Goethe-University, Frankfurt am Main. It was accepted as my doctoral thesis by the Faculty of Economics and Business Administration in October 2004. I am deeply indebted to the school, the department, and the staff working there for support and encouragement.

First of all I would like to thank my Ph.D. advisor, Prof. Christian Schlag. He taught me much of the foundations on which the research of this thesis has been based. Despite his busy schedule, he always found time to provide many extremely useful suggestions and provided me with a friendly and safe atmosphere in which to grow professionally. He has granted me broad freedom to pursue all of my ideas, and I am grateful for his show of confidence.

I would also like to thank the committee members Prof. Raimond Maurer, Prof. Reinhard H. Schmidt, and Prof. Dieter Nautz for reviewing my thesis and for their interest in my work.

My special thanks go to Dr. Angelika Esser without whose mathematical support I would not have got this far. Angelika was my officemate for three years and our close neighborhood and friendship resulted in two joint papers, which in turn were taken as Chapters 1 and 3 of this thesis.

Thanks are also due to Dr. Nicole Branger and Dr. Christoph Benkert for being a source of constant inspiration and for sharing their knowledge and ideas with me. I thank Micheal Belledin for teaching me how to program in C and Micong Klimes for her patience when constructing the order books and for a number of muffins, each one a delicious masterpiece. I owe thanks to Dr. Matthias Birkner and Dr. Roderich Tumulka for providing valuable help with some of the mathematical background for this thesis.

I would also like to thank Raisa Beygelman, Eva Schneider, and Norman Seeger for the stimulating and pleasant working atmosphere and for taking over some of my tasks during the last weeks before my defense, and our secretary Sigrid Peschel for all the administrative help. Thanks go to Jeffrey Boys and Sabine Steiner for copyediting and proofreading my thesis.

This book is about liquidity in financial markets and it would not exist without the liquidity that was generously provided by DFG, the German Research Foundation. Financial support by DZ Bank-Stiftung for the publication of this monograph is gratefully acknowledged. Furthermore, I am grateful to the Trading Surveillance Office of Deutsche Börse AG for providing an excellent limit order book data set.

Finally, my warmest thanks go to my dear parents for all the inexhaustible support and good advice over the years. They have been instrumental in furthering my education and helping me to get to this point.

Burkart Mönch, Frankfurt am Main   February 2005

Contents

List of Figures

List of Tables

Introduction

The Area of Research and the Object of Investigation

In this thesis we will investigate trading strategies in illiquid markets from a market microstructure perspective. Market microstructure is the academic term for the branch of financial economics that investigates trading and the organization of security markets, see, e.g., Harris (2002).

Historically, exchanges evolved as a location, where those interested in buying or selling securities could meet physically to transact. Thus, traditionally security trading was organized on exchange floors, where so-called dealers arranged all trades and provided liquidity by quoting prices at which they were willing buy or sell. Consequently, the initial surge of the market microstructure literature focused predominantly on this type of market design, which is often referred to as quote-driven.

Nowadays, the interest is shifting towards order-driven markets. Beginning with the Toronto Stock Exchange in the mid 1970s and increasing in frequency and scope, this market structure has emerged as the preeminent form of security trading worldwide. In order-driven markets, exchanges arrange trades by matching public orders, often by employing automatic execution systems.

A major difference between a quote-driven and an order-driven market arises from the transparency pre- and post-trade. The pre-trade transparency concerns the question whether the order book is visible to the keeper only, or whether it is open to the public. The post-trade transparency concerns the issue whether the details of recent trades are immediately reported to the public or not. Traditionally, most of the quote-driven markets were opaque, i.e. they offered only a low pre- and post-trade transparency, since the knowledge of the order flow could expose market makers to undue risk as they wanted to unwind positions. However, nowadays many order-driven markets, which are operated electronically, provide an order book that is completely or partially open to the public. In this thesis we will investigate the question of how investors can exploit the pre-trade transparency in order to derive optimal trading strategies. By choosing a market or a limit order, selecting a limit price, and by eventually specifying further contingencies under which circumstances the order should be executed, a market participant enjoys access to a wide range of strategies that trade off execution certainty against the expected execution price. Specifically, we will address the following four issues:

- Whether to break up orders.

- How to time order submissions.

- How to chose an reasonable limit.

- Whether to restrict the display of limit orders.

Our analysis will focus on large institutional investors (like insurance companies or pension funds) whose transaction sizes in a single stock represent a significant fraction of this stock's daily trading volume. These are particularly interesting, since acquiring or unwinding large positions in one security can incur significant costs which directly influence the return on the investment.

Imagine, for example a market order. Small market orders can usually be executed with little or no effect on prices and they pay one-half of the bid–ask spread for the opportunity to immediately trade. However, large market orders are more difficult to execute than small ones as they run through the order book until they are completely filled. The effect of executing a market order on prices is called price impact. It increases with the volume of the market order. Often, the price impact is the most significant cost of trading a large order. Hence, a large trader who uses a market order pays significantly more for immediacy than a retail investor who just wants to trade a few shares. Note that the price impact of a certain market order depends crucially on current market conditions. Since market conditions change quickly and unpredictably, traders who will use market orders at some point in the future cannot be sure about the prices they will receive. We will introduce an innovative liquidity derivative that offers protection against varying price impacts for a certain period of time.

Instead of using a market order, the large trader might want to employ a limit order. Investors who submit limit orders offer liquidity. Their orders give others the ability to trade when they want to trade. The originator of a limit order should keep in mind that displaying large orders in an open book may reveal his or her motives for trading. For small volumes, this fact may be negligible. However, institutional investors often break up large limit orders into smaller packages if they fear that showing their full sizes would cause the market to move away from them.

Iceberg orders facilitate these trading practices by executing such business automatically in the order book. They allow market participants to submit an order with only a certain portion of the order publicly disclosed – just as the iceberg's peak is the only visible portion of a huge mass of ice. Once the disclosed volume of an iceberg order has received a complete fill, a new peak appears in the book, with the volume equal to the initially disclosed amount. This loop is repeated until the whole iceberg is completely satisfied. At this point you may already suspect that the use of iceberg orders also comes at a cost, otherwise everybody would use them exclusively instead of limit orders and the order book would be completely opaque. In order to discover the differences between pure limit and iceberg orders one needs to consider the way orders are queued and processed in the order book. In almost any case this is governed by the price–time–priority rule. Price priority gives precedence to the lowest sell and the highest buy orders over all other orders stored on the same side of the book. Time priority means that among orders at the same price, the order placed earliest takes precedence. To control for time priority every order receives a time stamp when it appears in the book. This implies that a limit order enjoys a better overall time priority, as it has just one initial time stamp than an iceberg order where every peak is given a new time stamp when it shows up in the book.

Note that a limit order is basically a special case of an iceberg order where the peak size coincides with the overall order volume.

So far we have discussed the general area of research and we have presented the object of investigation: a large investor who is trading in an order-driven market that provides pre-trade access to order book data. Furthermore, we have introduced the basic order types available to the large trader: market, limit, and iceberg orders. There follows a brief overview of issues that will be analyzed in the next three chapters.

Outline of the Thesis

In the first chapter we model the interaction between the trading activities of a large investor, the stock price, and liquidity. Our framework generalizes the constant liquidity model of Frey (2000), and an extension by Frey and Patie (2001), where liquidity is made a deterministic function of the stock price by introducing a stochastic liquidity factor. This innovation has two implications. First, we can analyze trading strategies for the large investor that are affected by changes in market depth. Second, the sensitivity of stock prices to the trading strategy of the large investor can vary due to changes in liquidity. The main features of our model are demonstrated using Monte

Carlo simulation for different scenarios. The flexibility of our framework is illustrated by an application that deals with the pricing of a liquidity derivative. The claim under consideration compensates a large investor who follows a stop loss strategy for the liquidity risk that is associated with a stop loss order. The payoff of this claim can be best described by comparing a stop loss strategy in different regimes of liquidity.

If the market is perfectly liquid, as in the setup of Black and Scholes (1973), the investor will always receive the stop loss price for the asset. However, if the market becomes illiquid, the trader will receive a price less than the stop loss limit. If the degree of the illiquidity in the market does not change, as in Frey (2000), or can be interpreted as a function of the asset price, as in Frey and Patie (2001), this discount due to illiquidity is deterministic. This in turn means that the investor can adjust the stop loss limit in advance, so that he or she will always receive a deterministic amount for the assets. However, if order book depth is stochastic, the large investor faces liquidity risk.

The liquidity derivative matures when the stock price falls below the stop loss limit for the first time and then pays the price difference between the asset price immediately before and after the execution of the stop loss order. The setup to price the liquidity derivative is calibrated for one example using real-world limit order book data to give an impression of the magnitude of the liquidity effect. A number of empirical studies provide strong evidence that investors care about liquidity risk, which implies that this type of risk is priced into asset returns. Since liquidity is a non-traded risk factor the market price of liquidity risk appears in the pricing formula of our liquidity derivative. Similar to models with stochastic volatility it has to be estimated from other traded instruments that are exposed to this source of risk. We present a pragmatic approach to determine the market price of liquidity risk from traded European put options.

In the second chapter we will present a new pragmatic approach to determine optimal liquidation strategies if an investor uses market orders to unwind large security positions in an illiquid market. To keep the setup tractable we take a step backward and assume a deterministic relationship between the volume of a market order and the resulting price impact. Under the assumption that the liquidation horizon is given exogenously, the investor maximizes an objective function that considers the expected liquidation revenues and the respective standard deviation. The basic structure of our setup is closely related to approaches proposed by Bertsimas and Lo (1998), Almgren and Chriss (2000), Hisata and Yamai (2000), and Dubil (2002). While these authors focus mainly on theoretical aspects with the intention of deriving closed-form solutions for special types of market impact functions, we will propose a framework that is able to capture important empirical phenomena in the stock market. Specifically, the model contains:

- a U-shape for intraday stock market liquidity,

- power price impact functions,

- periods that allow the order book to be rebuilt, as boundary conditions for the time between subsequent trades.

Furthermore, it allows us to incorporate fixed transaction costs, as fixed charges incurred by the exchange or opportunity costs for handling the transactions in the front and back offices. The new model is very flexible since it allows for liquidation intervals of varying length and forgoes the assumption of a constant speed of trading. Examples with real-world order book data demonstrate how the setup can be implemented numerically and provide a deeper insight into relevant properties of the model.

As already discussed briefly in the previous section, market participants with large orders to execute are often reluctant to expose these to an open order book in their entirety in order to avoid a potential adverse market impact. In these situations investors often use iceberg orders.

In the third chapter we analyze the rationale for the use of this order type by assessing the costs and benefits of this trading instrument. At least to our knowledge, this is the first analytical approach that allows the determination of the optimal limit and the optimal peak size of an iceberg order. We assume that the investor who has to liquidate a large position in a stock within a finite time horizon follows a static strategy, i.e. once the limit and the peak size of the iceberg order are chosen, the trader sticks to this strategy over a fixed period. In our setup we will balance the relative advantage of hiding the actual order volume against the overall time priority of the iceberg order that deteriorates the smaller the peak size that is chosen.

Note that unless an iceberg or a limit order is immediately executable, i.e. the limit is set so aggressively that it is actually a market order, the probability of receiving a complete fill within a finite time horizon is strictly smaller than one. We propose two different approaches to incorporate the execution risk into our model. The first one assumes that the investor is forced to trade the remaining shares with a market order if the iceberg order fails to receive a complete fill within the prespecified time horizon. This setup is referred to as the self-contained approach. The second framework considers the execution probability as a boundary condition, i.e. only those combinations of peak size and limit are admissible that ensure a certain execution probability within a prespecified time horizon. This model is referred to as the open approach.

Once again we use a clinical order book data sample to explore important features of the model.

References

Almgren, R. and N. Chriss (2000): Optimal execution of portfolio transactions, *Journal of Risk* 3 (Winter 2000/2001), 5–39.

Bertsimas, D. and A. W. Lo (1998): Optimal control of execution, *Journal of Financial Markets* 1, 1–50.

Black, F. and M. Scholes (1973): The pricing of options and corporate liabilities, *Journal of Political Economy* 81, 637–54.

Dubil, R. (2002): Optimal liquidation of large security holdings in thin markets, University of Connecticut, Storrs, USA.

Frey, R. (2000): Market illiquidity as a source of model risk in dynamic hedging, in *Model Risk*, ed. by R. Gibson, RISK Publications, London, 125–36.

Frey, R. and P. Patie (2001): Risk management for derivatives in illiquid markets: a simulation study, RiskLab, ETH-Zentrum, Zürich, Switzerland.

Harris, L. (2002): *Trading and Exchanges*, Oxford University Press.

Hisata, Y. and Y. Yamai (2000): Research toward the practical application of liquidity risk evaluation methods, *Monetary and Economic Studies*, December, 83–128.

1 Modeling Feedback Effects with Stochastic Liquidity

1.1 Introduction

Aspects of market liquidity include the time involved in acquiring or liquidating a position and the price impact of this action. In this chapter we focus on the second issue. For large institutional investors in particular, many of the existing pricing models are only of limited use since they assume perfectly elastic supply and demand functions for the asset under consideration. This assumption is often violated if the trading activity of a large investor accounts for a significant fraction of the overall turnover in an asset. In this case, ignoring liquidity issues can result in a serious underestimation of the risk that is inherent in a certain investment strategy.

We present a framework that incorporates the liquidity risk arising for a large investor, whose trading volume cannot be absorbed by the market without a significant price change. Our model has two main ingredients. On the one hand, the stock price process is influenced by the trading activity of the large investor, whereas the impact of the trading strategy on the stock price is modeled using a stochastic liquidity (henceforth SL) factor. On the other hand, the stock price and the liquidity factor can have an impact on the trading strategy of the large investor.

In principle one could describe this scenario in two ways. First, one can build an equilibrium setup to explain the machinery of the market. Such an approach is sensible if one intends to analyze the motivation for trading or to investigate strategies that the large investor can use in order to exploit the power to move prices in a certain direction, see, e.g., Kyle (1985). However, for pricing purposes such complex frameworks are often unsuitable, as they are difficult to calibrate. In this chapter we follow a second approach by directly modeling the asset price dynamics that result if the large trader follows a certain trading strategy. We assume that the liquidity in the market is given exogenously.

There is a growing theoretical literature that investigates the interaction of liquidity and trading strategies of large investors. Part of this literature considers optimal liquidation strategies for large portfolios. Dubil (2002), Hisata and Yamai (2000), Almgren and Chriss (2000), and Bertsimas and Lo (1998) are just a few examples. Another branch of this literature investigates how large traders can manipulate stock prices. Jarrow (1992), Allen and Gale (1992), and Schönbucher and Wilmott (2000) can be mentioned in this line. Recent research focuses more and more on the modeling and hedging aspects that are introduced by illiquidity and the presence of one or more large traders. Cvitanić and Ma (1996), Cuoco and Cvitanić (1998), Sircar and Papanicolaou (1998), Frey (1998, 2000), Schönbucher and Wilmott (2000), Kampovsky and Trautmann (2000), Frey and Patie (2001), Liu and Yong (2004), and Bank and Baum (2002) are some prominent examples.

Our approach generalizes both the model of Frey (2000), where liquidity is constant, and an extension by Frey and Patie (2001), where liquidity is a deterministic function of the stock price. Modeling liquidity as a stochastic factor first of all enables us to incorporate random changes in market depth. Furthermore, we can significantly generalize existing models by introducing the concept of liquidity feedback effects. The presence of liquidity feedback effects implies that (i) trading strategies of large investors are affected by the degree of illiquidity and (ii) the sensitivity of stock prices to the trading strategies of large investors can vary due to changes in liquidity.

The objective of our model is to provide a large investor with a flexible framework that allows us to evaluate the liquidity risk associated with different types of trading strategies. We use a stop loss order as an example and analyze the effects of market liquidity for this type of trading strategy in detail. To get a flavor of the problem, imagine a pension fund. The fund management has to limit the downside risk of the fund, and in order to do so, assume that it follows a simple stop loss strategy. When the price of the security falls below a certain level the position in this security is liquidated completely and immediately by placing a market order. If the market is perfectly liquid the fund will always receive the stop loss price for the asset. However, if the market becomes illiquid, the investor will receive a price less than the stop loss limit. If the degree of the illiquidity in the market does not change or can be interpreted as a function of the asset price, this discount due to illiquidity is deterministic. This means that the investor can adjust the stop loss limit in advance, so that he or she will always receive a deterministic amount for the assets. However, if order book depth is stochastic, a large investor faces liquidity risk.

We propose a liquidity derivative compensating for this liquidity discount. We show how the setup can be calibrated with market data and present a simple and innovative approach to determine the market price of liquidity risk from the prices of traded European plain vanilla put options.

The chapter is organized as follows: Section 1.2 summarizes the main ideas of Frey's model. Section 1.3 motivates the introduction of liquidity as an autonomous source of risk. In a clinical study we show that modeling liquidity as a constant parameter or a deterministic function of the stock does not seem appropriate. Our general framework is presented in Section 1.4. In Section 1.5 we derive the effective dynamics for the underlying asset with SL. We use these results to compare the stock price dynamics in our SL model with the benchmark cases of geometric Brownian motion (as in the case of Black and Scholes (1973) (henceforth BS)) and constant liquidity (as in Frey (2000)). In Section 1.6 we exemplify the effects of SL by simulating sample paths of the asset price. Section 1.7 describes the general setup for a liquidity derivative, to illustrate how the SL model can be used to build pricing tools. The model is calibrated for an example using real-world limit order book data so that one gets an impression of the magnitude of the liquidity effect. Furthermore, we present an innovative approach to estimate the market price of liquidity risk from the prices of European put options. The chapter concludes in Section 1.8 with a brief summary and a discussion of issues for further research.

1.2   The Deterministic Liquidity Model

We now describe the setup proposed by Frey (2000). Assume there exists a risky asset S (the stock) and a risk-free investment earning a zero interest rate (the bond). There are no liquidity effects on the bond; only the underlying asset S is affected by this source of risk. Furthermore, there is a single large investor whose trading strategy influences the price process of the underlying asset. The risky asset follows the stochastic differential equation

$$\mathrm{d}S_t = \sigma S_{t-}\mathrm{d}W_t^S + \rho S_{t-}\mathrm{d}\phi_t^+,$$

where ϕ denotes the trading strategy of the large investor, i.e. the number of stocks held by him or her. ϕ^+ denotes the right-continuous version of ϕ, and $\rho \geq 0$ is a constant liquidity parameter. An increase in ρ means a declining liquidity in the market. For $\rho = 0$ the model represents the standard BS setup with zero drift. The quantity $1/(\rho S)$ is called the market depth, i.e. the order size that moves the price by one unit. Furthermore, we need the assumption that $\rho\,\mathrm{d}\phi_t^+ > -1$ in order to ensure non-negativity of the asset price.

Frey (2000) discusses the impact of the trading strategy on the price process for the case of a smooth strategy $(\phi \equiv \phi(t, S) \in C^{1,2})$[1]. The partial deriv-

---

[1]$C^{1,2}$ denotes the set of functions in two variables that are once continuously differentiable in the first variable and twice continuously differentiable in the second variable.

atives of ϕ are denoted by subscripts for ease of notation. This yields the effective dynamics for the underlying asset,

$$dS_t = b(t, S)dt + Sv(t, S)dW_t^S,$$

where

$$v(t, S) = \frac{\sigma}{1 - \rho S \phi_S}$$

$$b(t, S) = \frac{\rho S}{1 - \rho S \phi_S} \left(\phi_t + \frac{1}{2} \phi_{SS} S^2 v^2 \right)$$

assuming $\rho S \phi_S < 1$. Note that volatility has changed from σ to $\sigma/(1 - \rho S \phi_S)$ compared with a perfectly liquid market.

As discussed in Frey (2000) there are two basic types of trading strategies: On the one hand, the large investor can use a *positive feedback* strategy, i.e. $\phi_S > 0$. That means he or she buys the risky asset when the price is increasing, and he or she sells when the price is declining, thus reinforcing the effect of rising or falling prices. For example, in a standard BS model one would use such a strategy to duplicate a convex payoff like a long call. On the other hand, the large trader can employ a *contrarian feedback* strategy, i.e. $\phi_S < 0$, which means buying stocks when prices drop and vice versa. This would be the strategy used to duplicate a concave payoff, like a short call.

The basic model of Frey (2000) is extended in the paper of Frey and Patie (2001) by introducing a deterministic liquidity function $\rho(S)$. The extension does not change the fundamental properties of the model, since liquidity is not an autonomous source of risk as it is perfectly correlated with the risky asset.

1.3 Is Market Liquidity Indeed Stochastic?

In this section we briefly discuss whether variations in liquidity can be explained empirically by variations in the asset price or whether they can be represented as proposed in Frey and Patie (2001). If this approach does not work and if liquidity turns out not to be constant, we take this as an indication that liquidity should be modeled as an autonomous source of risk.

From an economic point of view one may argue that investors have different motives for submitting orders to the stock market. On the one hand, there may be a close relationship between the trading activity and the dynamics of the stock prices, for example if investors follow feedback strategies or trade for speculative reasons. On the other hand, so-called noise traders buy and sell assets to invest cash not needed for consumption or to meet cash needs in unforeseen situations. Imagine that a lucky retail investor has won the national lottery and now wants to buy stocks, or that an insurance company has to sell shares after a major damaging event to compensate clients for losses suffered. Individual liquidity shocks that occur independently of the stock price dynamics induce stochastic changes in market liquidity. As a theoretical reference we can take, for example, Ericsson and Renault (2003). The authors explicitly model individual liquidity shocks to investors in a market for defaultable bonds and investigate optimal liquidation strategies if liquidity is stochastic.

From an empirical perspective one can state that a growing branch of literature provides empirical evidence that market liquidity exhibits an intraday U-shaped or a J-shaped pattern, which is remarkably stable over time. A review of related literature can be found in Coughenour and Shastri (1999) or in Ranaldo (2000). For a discussion of a functional form that is able to reproduce an intraday U-shape pattern see, for example, Mönch (2003).

However, the intraday patterns can explain variations in market liquidity only to a certain extent, otherwise a time series of daily data should not exhibit any variations in market liquidity. We analyzed daily limit order book data collected at 12.00 a.m. for every trading day from January 3 to March 28, 2002 for Medion AG, which was one of the most heavily traded shares at Neuer Markt, the former market segment for growing technology companies at the German stock exchange up to spring 2003. Figure 1.1 shows a scatterplot for best bid prices S and percentage liquidity discounts ρ when

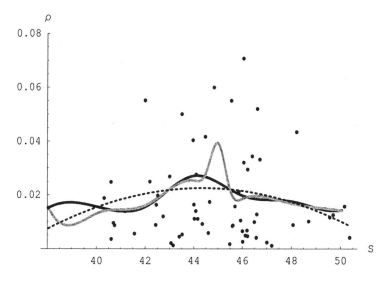

Fig. 1.1: Scatterplot for best bid prices S and percentage liquidity discounts ρ
if 5,000 Medion shares are sold in a single trade (daily data collected
at 12.00 a.m. CET for every trading day from January 3 to March 28,
2002). The dotted line represents the fitted function as proposed by
Frey and Patie (2001). The dark solid line shows a fitted cubic spline
for 6 intervals (5 interior knots). The light solid line represents a fitted
cubic spline for 10 intervals (9 interior knots).

5,000 Medion shares are to be sold in a single trade. At first sight it is evident
that daily liquidity discounts are not constant over time. Furthermore, one
can state that there is no significant relationship between the stock price and
the relative liquidity discounts. There is no clear pattern that would motivate
a certain functional form between asset price and liquidity risk. For example,
the fit of the function proposed in Frey and Patie (2001) given by

$$\rho\left(S\right) = \rho^{\text{const}} \cdot \left[1 - \left(S - S_0\right)^2 \left(a_1 I_{\{S \leq S_0\}} + a_2 I_{\{S > S_0\}}\right)\right]$$

provides an R^2 of only 0.03. The fitted function is plotted as a dotted line
in Figure 1.1. One may argue that other specifications of the function might
better explain variations in market liquidity. However, even cubic splines
that allow for a high degree of flexibility cannot adequately represent the

data. For example, for 6 intervals (5 interior knots) we obtain an R^2 of 0.05, for 10 intervals (9 interior knots) the R^2 is 0.06. The calibrated splines are plotted as a dark and a light solid line in Figure 1.1. Of course, one could increase the number of knots excessively to improve the fit of the underlying dataset. In the limit all data points may be taken as spline knots. In this case the spline interpolates the data points. However, such a model is not robust concerning the time window of the sample data used for calibration. Thus, its use would be questionable from an economic point of view.

1.4 The Stochastic Liquidity Model

In our model the underlying asset price process is assumed to follow the stochastic differential equation

$$dS_t = \mu_t S_{t-}dt + \sigma S_{t-}dW_t^S + \rho_t S_{t-}d\phi_t^+. \tag{1.1}$$

In order to be more general than Frey (2000) we relax the assumption of zero drift and interest rate. Furthermore, we now assume that the liquidity ρ follows a stochastic process with dynamics given by

$$d\rho_t = \beta(t,\rho)dt + \nu(t,\rho)dW_t^\rho \tag{1.2}$$

with the correlation specification

$$dW_t^S \bullet dW_t^\rho = \gamma dt. \tag{1.3}$$

A sensible choice for the stochastics of ρ could be a mean-reversion process with a natural long-run level of liquidity in the market. We further assume that the process stays strictly positive for $\rho_0 > 0$. This restricts the choices for the volatility function $\nu(t,\rho)$. For example, one might use functions of the type $\nu(t,\rho) = \zeta\sqrt{\rho}$.

Rewriting the above dynamics using a Cholesky decomposition of the covariance matrix of dS_t and $d\rho_t$ we obtain

$$dS_t = \mu_t S_{t-}dt + \sigma S_{t-}dW_t + \rho_t S_{t-}d\phi_t^+ \tag{1.4}$$

$$d\rho_t = \beta(t,\rho)dt + \nu(t,\rho)\gamma dW_t + \nu(t,\rho)\sqrt{1-\gamma^2}d\bar{W}_t \tag{1.5}$$

with a two-dimensional standard Brownian motion $(W_t, \bar{W}_t)$. We now assume that ϕ – the number of shares held by the large investor – does not only depend on S and t, as is the case in the model with deterministic liquidity, but also on the SL factor ρ. The effects on the trading strategy are now twofold: First, ϕ is influenced by changes in S. Second, ϕ varies with changing liquidity. The impact of S on ϕ can be modeled by the two basic types of trading strategies. For a positive feedback strategy, ϕ is an increasing function of S for all ρ; for a contrarian feedback strategy, ϕ is a decreasing function of S for all ρ. To characterize the impact of ρ on ϕ consider the following scenario. The more illiquid the market, the fewer shares the large trader will hold due to external or internal regulations, no matter whether a positive or contrarian feedback strategy is considered. Thus, a reasonable choice would be a decreasing absolute ϕ-value with respect to ρ (for all S).

1.5 The Effective Price Process in the Stochastic Liquidity Model

In this section we derive the effective price process for the SL model and analyze it in detail. Consider a smooth trading strategy $(\phi \equiv \phi(t, S_t, \rho_t) \in C^{1,2,2})^2$. An application of Itô's formula leads to the following proposition.

Proposition 1.1 (Dynamics of the State Variables) *Suppose the trading strategy of the large trader is given by $\phi(t, S, \rho) \in C^{1,2,2}$. Then, under the assumption $\rho S \phi_S < 1$ for any point in time, the solution to the system of stochastic differential equations (1.4) and (1.5) satisfies*

$$dS_t = b(t, S, \rho)dt + v(t, S, \rho)SdW_t + \bar{v}(t, S, \rho)Sd\bar{W}_t \qquad (1.6)$$

$$d\rho_t = \beta(t, \rho)dt + \nu(t, \rho)\gamma dW_t + \nu(t, \rho)\sqrt{1 - \gamma^2}d\bar{W}_t, \qquad (1.7)$$

---

[2]$C^{1,2,2}$ denotes the set of functions in three variables that are once continuously differentiable in the first variable and twice continuously differentiable in the second and third variable.

where

$$v(t, S, \rho) = \frac{\sigma}{1 - \rho S \phi_S} + \gamma \frac{\rho \phi_\rho \nu}{1 - \rho S \phi_S} \tag{1.8}$$

$$\bar{v}(t, S, \rho) = \sqrt{1 - \gamma^2} \frac{\rho \phi_\rho \nu}{1 - \rho S \phi_S} \tag{1.9}$$

$$b(t, S, \rho) = \frac{\rho S}{1 - \rho S \phi_S} \left[\left(\frac{\mu_t}{\rho} + \phi_t + \beta(\rho_t, t)\phi_\rho + \frac{1}{2}\phi_{\rho\rho}\nu^2 \right) \tag{1.10} \right.$$
$$\left. + \frac{1}{2}\phi_{SS}S^2(v^2 + \bar{v}^2) + \nu S \phi_{\rho S} \left(\gamma v + \sqrt{1 - \gamma^2}\bar{v} \right) \right].$$

The proof is given in Appendix 1.9.1.

In the case of $\rho \equiv 0$ or $\phi \equiv$ constant we are in the classical BS scenario with drift μ_t and constant volatility σ. If $\rho \neq 0$ and $\phi \neq 0$ the trading strategy of the large trader has an effect on the instantaneous volatilities v and $\bar{v}$, as well as on the total volatility and the correlation between the two processes.

In the special case where liquidity has no impact on the strategy of the large trader, i.e. $\phi_\rho \equiv 0$, we get close to the scenario of Frey (2000). Then, $\bar{v}$ and the second summand of v will vanish and feedback effects are only incorporated due to the term $\rho S \phi_S$. However, since ρ is stochastic in our model, the sensitivity of the stock price w.r.t. the trading activity of the large investor will now vary, in contrast to the deterministic liquidity model.

The analysis is more complex for $\phi_\rho \neq 0$. We assume $\rho > 0$, in order to discuss how variations in liquidity influence the trading strategy of the large investor. Assume that when illiquidity increases, the large trader has to reduce the position in the stock. Thus, the investor sells shares if he or she has a long position, or he or she buys back shares if a short position is considered. In the first case, ϕ is monotonically decreasing in ρ, starting with a positive ϕ. In the latter case, ϕ is monotonically increasing in ρ, starting with a negative ϕ. Thus, ϕ is approaching zero in absolute value (i.e. the large investor has closed the position in the stock almost completely) as ρ tends to infinity.

We assume a positive ϕ in the following so that ϕ_ρ should be negative, no matter if a positive feedback or a contrarian trading strategy is considered.

From (1.8) we can see that the sign of $v(t, S, \rho)$ depends on γ. For $\gamma \equiv 0$ the value of v is the same as in the deterministic liquidity model. Nevertheless, in the SL model we have an additional volatility parameter $\bar{v}$ contributing to total volatility. The parameter $\bar{v}$ is negative since ϕ_ρ is negative. The instantaneous quadratic variation $d[S]$ is given by (see Appendix 1.9.1)

$$d[S]_t = S^2 \left(\frac{\sigma^2 + \nu^2 \rho^2 \phi_\rho^2 + 2\gamma\sigma\nu\rho\phi_\rho}{(1 - \rho S \phi_S)^2} \right) dt$$

so that the total instantaneous volatility v_{tot} is equal to

$$v_{tot} = \sqrt{v^2 + \bar{v}^2} = \frac{\sqrt{\sigma^2 + \nu^2 \rho^2 \phi_\rho^2 + 2\gamma\sigma\nu\rho\phi_\rho}}{1 - \rho S \phi_S}. \qquad (1.11)$$

The instantaneous covariation of the two processes $d[S, \rho]$ is given by

$$\begin{aligned} d[S, \rho]_t &= S\nu \left(\gamma v + \sqrt{1 - \gamma^2} \bar{v} \right) dt \\ &= S\nu \left(\frac{\nu\rho\phi_\rho}{1 - \rho S \phi_S} + \gamma \frac{\sigma}{1 - \rho S \phi_S} \right) dt. \qquad (1.12) \end{aligned}$$

Thus, the instantaneous correlation η equals

$$\begin{aligned} \eta = \frac{d[S, \rho]}{\nu S v_{tot} dt} &= \frac{\gamma v + \sqrt{1 - \gamma^2} \bar{v}}{\sqrt{v^2 + \bar{v}^2}} \\ &= \frac{\nu\rho\phi_\rho + \gamma\sigma}{\sqrt{\sigma^2 + \nu^2 \rho^2 \phi_\rho^2 + 2\gamma\sigma\nu\rho\phi_\rho}}. \qquad (1.13) \end{aligned}$$

In order to compare the formal setup of Frey (2000) and Frey and Patie (2001) with the SL model, we analyze the volatility and correlation structure for different specifications of the respective liquidity-related parameters. An overview is given in Table 1.1.   For $\rho \equiv 0$ or $\phi \equiv$ constant, respectively, we are in the standard BS model with drift μ, resulting in $\bar{v} \equiv 0$ and $v = \sigma$. The first additional feature is included in Frey's approach where liquidity is represented by a constant ρ (i.e. $d\rho \equiv 0$) implying $\beta \equiv \nu \equiv 0$. This yields $\bar{v} \equiv 0$ and

$$v \equiv v_{tot} = \frac{\sigma}{1 - \rho S \phi_S}.$$

The correlation parameter η must be zero in this scenario, since all terms containing the partial derivative ϕ_ρ vanish (so that the dependence of the strategy on ρ is of no interest for the effective stock price dynamics).

Table 1.1: Liquidity-related parameters in different models.

Model	γ	v	$\bar{v}$	v_{tot}	η
BS	0	σ	0	v	0
Frey	0	$\frac{\sigma}{1-\rho S \phi_S}$	0	v	0
Frey – Patie	1	$\frac{\sigma}{1-\rho S \phi_S}$	0	v	1
SL	0 (uncorr.)	$\frac{\sigma}{1-\rho S \phi_S}$	$\frac{v \rho \phi_\rho}{1-\rho S \phi_S}$	$\sqrt{v^2+\bar{v}^2} > v$	$\frac{\bar{v}}{\sqrt{v^2+\bar{v}^2}} < 0$
SL	$\neq 0$ (corr)	$\frac{\sigma + \gamma v \rho \phi_\rho}{1-\rho S \phi_S}$	$\frac{\sqrt{1-\gamma^2}\, v \rho \phi_\rho}{1-\rho S \phi_S}$	$\sqrt{v^2+\bar{v}^2}$	$\frac{\sqrt{1-\gamma^2}\,\bar{v}+\gamma v}{\sqrt{v^2+\bar{v}^2}}$

Now we take a closer look at the model of Frey and Patie (2001), where ρ is a deterministic function of S. Therefore, the dynamics of ρ are only driven by the first component W of the two-dimensional Brownian motion, which implies $\bar{v} \equiv 0$ and $\gamma = 1$. Since ϕ depends only on S, the volatility is the same as in the Frey setup. From (1.13) we can deduce that $\eta \equiv 1$ in this case. Thus, the approach of Frey and Patie (2001) is a special case of our general framework with $\gamma = 1$, allowing the coefficients in the dynamics of ρ to depend explicitly on S.

We now consider the SL model and start with the correlation structure.

1.5.1 Impact on Correlation

It is important to note that $\gamma = 0$ does not imply that the increments of the effective stock price process and the liquidity process are uncorrelated, as one can see in equation (1.13). In the SL setup the situation where $\eta = 0$ cannot be obtained for a deterministic choice of γ. Even for $\gamma = 0$, there is still some correlation η between dS and $d\rho$ induced by $\rho\phi_\rho$:

$$\eta = \frac{\nu\rho\phi_\rho}{\sqrt{\sigma^2 + \nu^2\rho^2\phi_\rho^2}} < 0.$$

The parameter η is negative since $\phi_\rho < 0$. This term ultimately models liquidity feedback effects. It vanishes for a trading strategy independent of ρ, i.e. for $\phi_\rho \equiv 0$. The negative value of η can be interpreted in the following way: If liquidity decreases over a longer period the trader will be forced to close the position, which will cause the stock price to drop. For $\gamma \neq 0$ the numerator of the correlation η in equation (1.13) carries an additional summand $\gamma\sigma$. For $\phi_\rho \equiv 0$ the correlation η between dS and $d\rho$ is equal to γ, the correlation between the increments of the components of the Brownian motion. Thus, the difference between γ and η is an exclusive result of the liquidity feedback effect.

Next, we discuss the impact of SL on volatility.

1.5.2 Impact on Volatility

First, consider the case $\gamma = 0$. Then, v is the same as in the deterministic liquidity model, but in our framework there is the additional volatility parameter

$$\bar{v} = \frac{\nu\rho\phi_\rho}{1 - \rho S\phi_S} < 0.$$

The variable $\bar{v}$ incorporates the liquidity feedback effect. It contributes to total volatility if and only if the trading strategy depends on the liquidity parameter ρ.

In this case total volatility increases, compared with the deterministic liquidity model, and we obtain a total volatility of

$$v_{tot} = \frac{\sqrt{\sigma^2 + \nu^2\rho^2\phi_\rho^2}}{1 - \rho S\phi_S}.$$

Here we see a key result of our approach: For $\gamma = 0$ the total volatility of the SL model is greater than the volatility in the constant liquidity setup, no matter which strategy is used.

For $\gamma \neq 0$ the direction of the changes in total volatility (see equation (1.11)) depends on the sign of γ. The expression for v contains an additional summand describing the liquidity feedback effect, and $\bar{v}$ is reduced in absolute values due to $\gamma \neq 0$ (see equations (1.8) and (1.9)). The total volatility in equation (1.11) is a decreasing function of γ, since $\phi_\rho < 0$. Thus, even for $\gamma \leq 0$ the total volatility is greater than the volatility in Frey's deterministic liquidity setup (2000). Only for large positive values of γ is it possible that the total volatility is lower than the volatility in Frey's (2000) model. This will be the case if

$$1 \geq \gamma > \frac{1}{2\sigma}\nu\rho|\phi_\rho|. \tag{1.14}$$

For a strategy ϕ independent of ρ the total volatility in the SL model is equal to the volatility in the constant liquidity model, again reflecting the lack of the liquidity feedback effect in this case.

Finally, we look at the two basic types of trading strategies with respect to S.

1.5.2.1 Positive Feedback Strategy

A positive feedback strategy, i.e. $\phi_S > 0$, leads to

$$v_{tot} > \sqrt{\sigma^2 + \rho^2\nu^2\phi_\rho^2 + 2\gamma\sigma\nu\rho\phi_\rho}.$$

The expression on the right-hand side is greater than or equal to σ if and only if

$$\gamma \leq \frac{1}{2\sigma}\nu\rho|\phi_\rho|. \tag{1.15}$$

Importantly, v_{tot} is greater than the BS volatility σ for $\gamma \leq 0$. This is similar to the result derived in Frey (2000).

1.5.2.2 Contrarian Feedback Strategy

A contrarian feedback strategy (i.e. $\phi_S < 0$) implies that the total volatility satisfies

$$v_{tot} < \sqrt{\sigma^2 + \rho^2 \nu^2 \phi_\rho^2 + 2\gamma\sigma\nu\rho\phi_\rho}$$

which is less or equal to σ if and only if (1.14) holds. Thus, for non-positive values of γ the instantaneous volatility in the SL model is not lower than the BS volatility in this case. Note that this is opposite to the result in Frey's (2000) model.

1.6 Numerical Results

In order to visualize the formal analysis of the previous section we present some simulation-based results. We compare a sample path of stock prices in the BS model with stock prices in the SL setting for both a positive and a contrarian feedback strategy.

1.6.1 Parameter Specification

In order to ensure non-negativity and stationary behavior of the liquidity process we specify the dynamics for ρ in equation (1.2) as a square-root process with a mean-reverting drift component (see Cox, Ingersoll, and Ross (1985) (henceforth CIR)):

$$\beta(t,\rho) = \kappa(\theta - \rho), \quad \nu(t,\rho) = \zeta\sqrt{\rho}.$$

To create paths of the underlying and the liquidity parameter, Monte Carlo simulation techniques are used. For a fixed realization of $(W_t, \bar{W}_t)$ we have used the dynamics of equations (1.4) and (1.5) for $S_0 = 80.0$, $\sigma = 0.1$ and

1. with $\rho \equiv 0$ in the BS setting and

2. for the SL setup: $\kappa = 0.35, \quad \theta = \rho_0 = 0.05, \quad \zeta = 0.2, \quad \gamma = 0.0$.

The stochastic processes are discretized using an Euler scheme with $N = 4000$ steps and time intervals of length $\Delta t = 1/360$.

Figure 1.2 shows the stock holdings of the large investor as a function of ρ and S for a positive (left graph) and a contrarian feedback strategy (right graph).

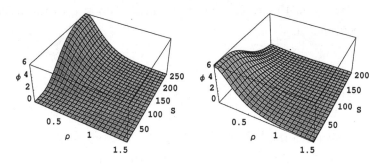

Fig. 1.2: Positive and contrarian feedback strategy.

The shape of the functions can be explained by the following intuition. If liquidity drops, i.e. ρ increases, the large trader is forced to sell shares. Therefore, ϕ is assumed to be monotonically decreasing in ρ for all S and for any feedback strategy, i.e. the derivative of ϕ with respect to ρ is negative. It seems reasonable to assume that for very small and for very large values of ρ the absolute value of ϕ_ρ is small. In the first case, the asset still has a sufficient market depth. In the latter case, the large trader has already sold almost all of his holdings in the stock. Thus, in both scenarios, the large trader adjusts the position in the stock by only a small amount.

In order to characterize the relationship between ϕ and S we have to distinguish between the positive feedback and the contrarian feedback strategy. In the first (second) case, the large trader buys (sells) assets as the stock price increases and sells (buys) when the stock price declines. Thus, ϕ is monotonically increasing (decreasing) in S for all ρ in the case of the positive (contrarian) feedback strategy. For very small and very large values of S

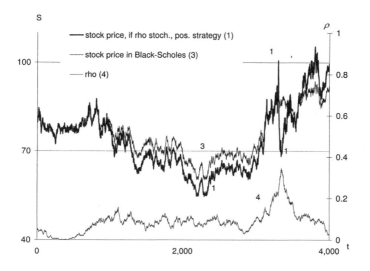

Fig. 1.3: Sample paths for the stock price in the BS model and the SL setup if the large trader follows a positive feedback strategy.

the changes of the stock holdings of the large trader are negligible when the asset prices vary (similar to the relationship between ρ and ϕ). However, for asset prices in between, the absolute value of ϕ_S increases when S increases and a positive (contrarian) feedback strategy is considered.

There are a variety of functional forms for $\phi = \phi(S, \rho)$ that are able to reproduce those features described above. We use the incomplete gamma function to model this scenario. The exact functional form for $\phi(S, \rho)$ can be found in Appendix 1.9.2.

1.6.2 Positive Feedback Strategy

In Figure 1.3 we compare stock price paths generated by the BS model with those produced by our SL model in the case of a positive feedback strategy.

As shown analytically in equation (1.15) one can see that for the given choice of parameters, the volatility of the stock price in the SL model is increased compared with BS. If the large investor follows a positive feedback strategy in a bullish market, the stock price dynamics in the SL model exceed the corresponding stock prices in a BS world. Rising stock prices motivate the large investor to buy additional stocks, which will cause the stock price to grow even further in an illiquid market. One can notice the opposite effect for a decreasing S since in this case the large investor wants to get rid of the holdings, which will accelerate the decline in the stock price.

The role of the liquidity parameter ρ is more subtle. In fact it can have two different implications: First, all else equal, the trader has to sell shares if they become more and more illiquid. Second, if ρ is very high, the stock becomes more volatile so that a large trader who follows a positive feedback strategy can cause the stock price to rise to tremendously high values in bullish markets. However, when illiquidity exceeds a certain threshold ($\rho \approx 0.3$ in Figure 1.3), the large trader is forced to close the position, and the market collapses. These features, which distinguish the SL model from the BS and the deterministic liquidity model, become evident around the time step 3,500 in Figure 1.3.

When ρ approaches zero the stock price in the SL model runs parallel to the stock price in the BS model, as one can observe around time step 500.

1.6.3   Contrarian Feedback Strategy

In Figure 1.4 we contrast stock prices simulated in the BS model with those generated in the SL model in the case of a contrarian feedback strategy.

In general, volatility is reduced compared with BS. However, there are also exceptions, as one can observe around time step 3,500. When the asset becomes very illiquid the trading activities of the large trader can dominate the stock price dynamics and so have a destabilizing effect. Again, this is a unique feature of the SL setup that cannot be reproduced in the constant liquidity framework.

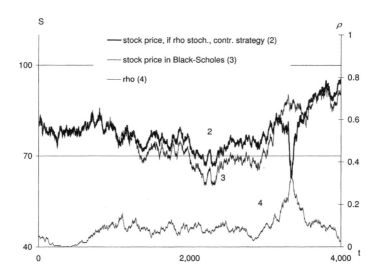

Fig. 1.4: Contrarian feedback strategy versus Black–Scholes.

1.7   Application: Liquidity Derivative

After having introduced the general framework of our model, we now present
an application that deals with the pricing of a liquidity derivative on an
underlying asset that is traded in an illiquid market. To ease the exposition
we take a step backward and restrict the general framework so that the
trading strategy of the large investor depends only on the asset price S,
not on liquidity ρ. The derivative under consideration compensates the large
investor following a stop loss strategy for the liquidity discount if the stop loss
order is executed. We provide an example of such a derivative, the pricing
of which is based on our sample data from the German electronic limit order
book XETRA for the technology company Medion.

In contrast to retail investors whose individual trading volume is too small
to affect prices adversely, large traders have to consider liquidity aspects before an investment is made. If the degree of illiquidity in a certain market is
rapidly changing over time one may hesitate to invest substantial amounts,
although the risk-reward profile of the investment per se might be promis-

ing. A long position in the proposed liquidity derivative hedges against the liquidity risk and enables a large trader to act like a small retail investor. Currently such liquidity derivatives are not yet actively traded in the market. The lack of appropriate pricing tools and order book data needed for calibration may partly explain this situation. However, as competition among electronic trading platforms sharpens, many exchanges might provide clients with real-time access to data on market liquidity.

Trading strategies that limit the downside risk of a portfolio and that depend only on the asset price are still heavily used by institutional investors. Considering a simple stop loss strategy, however, one might argue that a large trade in the stock market is usually broken up into smaller packages to minimize the adverse impact on the overall transaction price. We nevertheless focus on this application since, despite its simplicity, it already incorporates a basic structure and can thus serve as a guideline to construct more elaborate pricing tools.

1.7.1 Pricing Formulas

Assume that ϕ is a stop loss trading strategy. The random variable $\tau := \inf\{t \mid S_t < \bar{S}\}$ denotes the stopping time when the underlying asset falls below a certain level $\bar{S}$ for the first time, implying $S_{\tau-} = \bar{S}$. The initial price of the underlying asset is assumed to be $S_0 > \bar{S}$. Up to the hitting time τ the large trader does not trade, and at τ he or she sells all assets. Assuming a constant, positive initial value $\phi_0 > 0$, this implies

$$d\phi_t^+ = \begin{cases} 0 & \text{for } t \neq \tau \\ \phi_t^+ - \phi_t = -\phi_0 & \text{for } t = \tau \end{cases}$$

and

$$\begin{aligned} dS_t &= \mu S_t dt + \sigma S_t dW_t^S \\ d\rho_t &= \beta(t, \rho)dt + \nu(t, \rho)d\bar{W}_t^\rho \end{aligned}$$

for $t < \tau$. Equivalently, using the uncorrelated Brownian motion $(W, \bar{W})$, this can be written as

$$
\begin{aligned}
dS_t &= \mu S_t dt + \sigma S_t dW_t \\
d\rho_t &= \beta(t, \rho)dt + \gamma \nu(t, \rho)dW_t + \sqrt{1 - \gamma^2}\nu(t, \rho)d\bar{W}_t
\end{aligned}
$$

for $t < \tau$. For derivative pricing we need the dynamics under a risk-neutral measure $\hat{P}$. They are given by

$$
\begin{aligned}
dS_t &= rS_t dt + \sigma S_t d\hat{W}_t & (1.16) \\
d\rho_t &= \beta^*(t, \rho)dt + \gamma \nu(t, \rho)d\hat{W}_t + \nu(t, \rho)\sqrt{1 - \gamma^2}d\hat{\bar{W}}_t & (1.17)
\end{aligned}
$$

for $t < \tau$, where β^* denotes the risk-adjusted liquidity drift. Using

$$
\begin{aligned}
d\hat{W}_t &= \lambda_t^{(S)}dt + dW_t \\
d\hat{\bar{W}}_t &= \lambda_t^{(\rho)}dt + d\bar{W}_t^\rho
\end{aligned}
$$

it holds as usual that $\lambda_t^{(S)} = (\mu - r)/\sigma$, while $\lambda_t^{(\rho)}$ can be chosen arbitrarily, so that we obtain

$$
\beta^*(t, \rho) = \beta(t, \rho) - \gamma \nu(t, \rho)\frac{\mu - r}{\sigma} - \sqrt{1 - \gamma^2}\lambda_t^{(\rho)}.
$$

Since liquidity is not traded, the market is incomplete. This leads to one degree of freedom for the market price of liquidity risk, which cannot be eliminated in this setup. The market price of liquidity risk shows up in the difference between dynamics under P and under $\hat{P}$. To fix the drift of the liquidity process under $\hat{P}$ a unique price of at least one derivative depending explicitly on liquidity is needed. We provide a theoretical discussion on how the market price of liquidity risk can be calculated from traded European options in Subsection 1.7.4. This approach represents a perfect analogy to a stochastic volatility model where the market price of volatility risk can only be computed from the prices of derivatives.

Let us now come back to the stopping time τ. By definition, the threshold is hit at $\tau-$. This yields a jump in the stock price at τ given by $S_{\tau-} = \bar{S}$ and the reduced price at τ:

$$
S_\tau = \bar{S}(1 - \rho_{\tau-}\phi_0) = \bar{S}(1 - \rho_\tau \phi_0),
$$

assuming that ρ is a continuous function in τ. We propose a derivative contract that compensates for the price difference between the reduced price at τ and the threshold $\bar{S}$ by paying $\bar{S}\rho_\tau\phi_0 I(\tau \leq T)$ at τ for one unit of the underlying asset.

To model a more realistic scenario we additionally introduce a floor $F \geq 0$, which represents a deductible for the investor. This means that if the price discount due to illiquidity is less than or equal to F, the derivative pays nothing. Consequently, the contract would only compensate for a critical liquidity discount. Thus, the payoff at τ for one unit of the underlying asset is

$$
\begin{aligned}
Z_\tau &= \max[(\bar{S}\phi_0\rho_\tau - F)I(\tau \leq T); 0] \\
&= (\bar{S}\phi_0\rho_\tau - F)I\left(\tau \leq T; \rho_\tau > \frac{F}{\bar{S}\phi_0}\right).
\end{aligned}
$$

The price of the contract is given by the following proposition:

Proposition 1.2 *Let* $\tau := \inf\{t \mid S_t < \bar{S}\}$. *Then the price at* $t = 0$ *of a derivative paying* $Z_\tau = \max[(\bar{S}\phi_0\rho_\tau - F)I(\tau \leq T); 0]$ *is given by*

$$
\begin{aligned}
Z_0 &= \hat{\mathbf{E}}\left[e^{-r\tau}Z_\tau\right] \\
&= \bar{S}\phi_0\hat{\mathbf{E}}\left[e^{-r\tau}\left(\rho_\tau - \frac{F}{\bar{S}\phi_0}\right)I\left(\tau \leq T; \rho_\tau > \frac{F}{\bar{S}\phi_0}\right)\right] \\
&= \bar{S}\phi_0\int_0^T\int_{F/\bar{S}\phi_0}^{\bar{\rho}} e^{-rt}\left(\rho - \frac{F}{\bar{S}\phi_0}\right)g(t, \rho)\mathrm{d}\rho\mathrm{d}t, \quad (1.18)
\end{aligned}
$$

where $\hat{\mathbf{E}}$ denotes the expectation under the risk-neutral measure and g represents the joint (risk-neutral) density of τ and ρ_τ. The parameter $\bar{\rho} \leq \infty$ denotes some upper bound for the liquidity process. In general, the price of the derivative cannot be calculated explicitly, but for some special cases we are able to derive semi-closed-form solutions of (1.18). One of these is presented below.

First of all, we need the distribution of the hitting time. From (1.16) we know that the process for S up to τ under a risk-neutral measure $\hat{\mathbf{P}}$ is a geometric Brownian motion such that the log of the process is an arithmetic Brownian motion. The distribution of the first hitting time of a Brownian motion with drift is well-known (see, for example, Borodin and Salminen (1996)). In this scenario we need to know the distribution of the first hitting time of $\ln \bar{S}$, starting at $\ln S_0 > \ln \bar{S}$ under a risk-neutral measure. This is given by

$$f_{\ln S_0, \ln \bar{s}}(t) = \frac{|\ln(\bar{S}/S_0)|}{\sqrt{2\pi t^3}} \exp\left(\frac{-(\ln(\bar{S}/S_0) - (r - \sigma^2/2)t)^2}{2t}\right).$$

The specification of the diffusion process for liquidity has remained arbitrary so far. In the following we consider some scenarios in which the price of the liquidity derivative can be calculated explicitly under certain assumptions on the liquidity process.

If liquidity is constant or a random variable uncorrelated with the underlying asset, equation (1.18) simplifies to

$$
\begin{aligned}
Z_0 &= \bar{S}\phi_0 \left(\hat{\mathbb{E}}\left[\rho\right] - \frac{F}{\bar{S}\phi_0}\right) \hat{\mathbb{E}}\left[e^{-r\tau} I(\tau \le T)\right] \\
&= \bar{S}\phi_0 \left(\hat{\mathbb{E}}\left[\rho\right] - \frac{F}{\bar{S}\phi_0}\right) \int_0^T e^{-rt} f_{\ln S_0, \ln \bar{s}}(t)\mathrm{d}t.
\end{aligned}
$$

If the Brownian motions exhibit zero correlation, i.e. $\gamma = 0$, the hitting time τ and the process ρ are independent. In this case we can calculate the expectation, taking the product of the corresponding risk-neutral densities, i.e. $g(t, \rho) = f(t)h_t(\rho)$, where h_t denotes the risk-neutral density of the process ρ at time t, and we obtain

$$
\begin{aligned}
Z_0 &= \bar{S}\phi_0 \hat{\mathbb{E}}\left[e^{-r\tau}\rho_\tau I(\tau \le T)\right] \\
&= \bar{S}\phi_0 \int_0^T \left(\int_{F/\bar{S}\phi_0}^{\bar{\rho}} \left(\rho - \frac{F}{\bar{S}\phi_0}\right) h_t(\rho)\mathrm{d}\rho\right) e^{-rt} f_{\ln S_0, \ln \bar{s}}(t)\mathrm{d}t. \quad (1.19)
\end{aligned}
$$

For $F \equiv 0$ this simplifies to

$$Z_0 = \bar{S}\phi_0 \int_0^T \left(\hat{\mathbb{E}}[\rho_t]\right) e^{-rt} f_{\ln S_0, \ln \bar{s}}(t)\mathrm{d}t. \quad (1.20)$$

Now, we are free to choose an appropriate process for ρ. To our knowledge the question of which specification of the liquidity process is empirically adequate is still unanswered in the literature.

One may argue that a mean reversion process is a sensible choice, since there exists a natural level of liquidity in the market. Thus, we again assume a CIR process for liquidity given by the following specification of the risk-neutral parameters in equation (1.17):

$$\beta^*(t, \rho) := \kappa^*(\theta^* - \rho)$$
$$\nu(t, \rho) := \zeta\sqrt{\rho}.$$

The asterisks denote risk-neutral parameters. For zero correlation the dynamics are given by

$$d\rho_t = \kappa(\theta - \rho_t)dt + \zeta\sqrt{\rho_t}\,d\bar{W}_t \text{ under P and by}$$
$$d\rho_t = \kappa^*(\theta^* - \rho_t)dt + \zeta\sqrt{\rho_t}\,d\hat{\bar{W}}_t \text{ under } \hat{P}$$

with $d\hat{\bar{W}}_t = \sqrt{\rho_t}\lambda_t dt + d\bar{W}_t$. Here we set $\lambda_t^{(\rho)} = \sqrt{\rho_t}\lambda_t$ such that the liquidity process is again a CIR process after the change of measure. The parameters are then given by

$$\kappa^* = \kappa + \zeta\lambda, \quad \theta^* = \frac{\kappa}{\kappa^*}\theta. \tag{1.21}$$

In order to ensure a strictly positive liquidity path we have to impose the constraint

$$\theta^* > \frac{\zeta^2}{2\kappa^*}. \tag{1.22}$$

The risk-neutral probability density of ρ at t with initial value ρ_0 is given by

$$h(t, \rho; \rho_0) = ce^{-u-v}\left(\frac{v}{u}\right)^{q/2}\mathcal{I}_q(2\sqrt{uv})$$

where

$$c = \frac{2\kappa^*}{\zeta^2(1 - e^{-\kappa^* t})}$$
$$u = c\rho_0 e^{-\kappa^* t}$$
$$v = c\rho$$
$$q = \frac{2\kappa^*\theta^* - \zeta^2}{\zeta^2}.$$

$\mathcal{I}_z$ denotes the modified Bessel function of the first kind of order z, i.e.

$$\mathcal{I}_z(y) = \left(\frac{y}{2}\right)^z \sum_{n=0}^{\infty} \frac{(y^2/4)^n}{\Gamma(z+n+1)n!}.$$

Note that if the floor F is equal to zero the expectation of liquidity in equation (1.20) can be computed explicitly:

$$\mathbb{E}\left[\rho_t\right] = \rho_0 e^{-\kappa^* t} + \theta^* \left(1 - e^{-\kappa^* t}\right). \tag{1.23}$$

Hence, equation (1.20) simplifies to

$$Z_0 = \bar{S}\phi_0 \int_0^T \left(\rho_0 e^{-\kappa^* t} + \theta^* \left(1 - e^{-\kappa^* t}\right)\right) e^{-rt} f_{\ln S_0, \ln \bar{s}}(t) dt.$$

If one additionally assumes a zero market price of liquidity risk $(\theta^* = \theta)$, then the price of the liquidity derivative does not depend on the volatility parameter of the liquidity process ζ. However, as soon as these restrictive assumptions are relaxed, for example by considering a non-zero market price of liquidity risk $(\lambda^\rho \neq 0)$ or a non-linear payoff function $(F > 0)$, the volatility of the liquidity process has an impact on the price of the liquidity derivative. This property is shown below in Figures 1.6 and 1.7 in Subsection 1.7.2.

1.7.2 Example

As an example we use the limit order book data for Medion already introduced in Section 1.3 to calibrate the stochastic processes for the best bid price and liquidity. For numerical reasons we multiply the liquidity parameter by 5,000. Thus, it can be interpreted as the relative price difference between the best bid price and the average execution price if a *hypothetical* market order of 5,000 shares is executed.

Applying standard maximum likelihood estimation techniques and assuming a zero market price of liquidity risk, the following parameter estimates were obtained:

$$\begin{aligned} \sigma &= 0.48 & \theta^* &= 0.020188 \\ \kappa^* &= 248.12 & \zeta &= 3.1906347. \end{aligned}$$

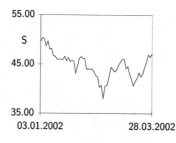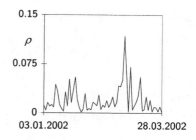

Fig. 1.5: Best bid prices and percentage liquidity discounts if 5,000 Medion shares are sold in a single trade (daily data at 12.00 a.m. CET).

The correlation between the best bid prices and the illiquidity discounts in the sample is -0.0354. Thus, assuming a correlation of zero seems to be sensible.

We assume that the derivative under consideration compensates for the price difference between the stop loss limit of €40.00 and the average execution price if a stop loss order of 5,000 shares is executed within the next month.[3] Since the initial liquidity parameter ρ_0 is scaled with 5,000 we have $\phi_0 = 1.0$.

Furthermore, we set $S_0 = 47.61$, $\rho_0 = 0.05$ and $r = 0.05$, and in the model with deterministic liquidity $\rho = \theta^* = 0.020188$. Now we are able to calculate prices for the liquidity derivative. In the model with constant liquidity one obtains a price of 0.465449, whereas in the model with SL the price for one derivative is 0.489582. Thus the price difference for one contract (in our example 5,000 derivatives) between the stochastic and the deterministic liquidity model would be €120.67.

1.7.3 Sensitivity Analysis

How does variation in the parameters affect the results in the SL model? Recall that the parameters of the CIR process have to meet the constraint $\theta^* > \zeta^2/(2\kappa^*)$ in order to ensure positivity of the liquidity process. When

---

[3]For the sake of simplicity we assume that a stop loss order is executed if the best bid price falls to the stop loss limit for the first time.

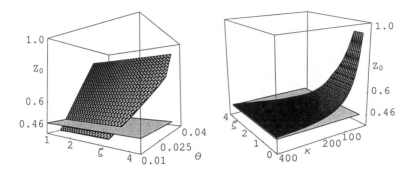

Fig. 1.6: Price of the liquidity derivative Z_0 as a function of the parameters ζ, θ^* (left) and ζ, κ^* (right). All other parameters are chosen in accordance with the example; we assume especially that $F = 0$ and $\lambda^{(\rho)} \equiv 0$. While the dark patterned surfaces represent Z_0 in the SL framework, the light surfaces represent Z_0 in the constant liquidity setup.

the boundary condition becomes binding, a reduction in κ^* or θ^* implies a reduction in ζ (for given θ^* or κ^*). Thus, the choice of parameters is restricted.

First we stay for a while in the slightly unrealistic world of the previous example and consider a contract that is linear in ρ_t, i.e. $F = 0$, and assume a zero market price of liquidity risk. Figure 1.6 illustrates how the price of the derivative Z_0 varies with either θ and ζ for fixed κ (left graph) or with κ and ζ for fixed θ (right graph) such that inequality (1.22) is always satisfied. The value of the derivative in the SL framework is represented by the dark patterned surfaces. The light surfaces represent the derivative in the constant liquidity setup, which serves as a benchmark. Under the given assumptions $F = 0$ and $\lambda^{(\rho)} = 0$, which imply $\theta^* = \theta$, $\kappa^* = \kappa$, the world is simple. Remember that in this case equation (1.19) simplifies to (1.20), where $\hat{\mathbb{E}}[\rho_t]$ is given by (1.23). In the SL setup the price of the liquidity derivative Z_0 is a linear increasing function of the long-term mean θ and an exponentially decreasing function of the mean reversion parameter κ.

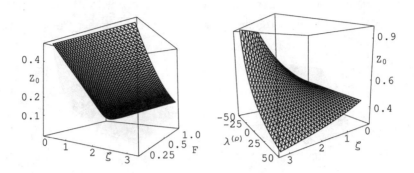

Fig. 1.7: Price of the liquidity derivative Z_0 as a function of the parameters ζ, F
(left) and ζ, $\lambda^{(\rho)}$ (right). All other parameters are chosen in accordance
with the example.

The volatility parameter ζ has no impact on the price of the derivative Z_0.
Reconsidering the definition of $\mathbf{E}\left[\rho_t\right]$ given by equation (1.23) one can state
that the stochastic and the deterministic liquidity setup give almost identical
results if ρ_0 and θ^* are close to $\rho^{const.}$ and if the product $\kappa^* t$ is sufficiently
large.

If the assumptions of the previous examples are relaxed the volatility parameter ζ has an impact on the price of the liquidity derivative Z_0. Figure 1.7
shows the price of the derivative Z_0 for combinations of ζ and F (left graph)
and of ζ and $\lambda^{(\rho)}$ (right graph) for fixed κ^* and θ^*.

When the floor F is increased, the derivative becomes less expensive. In this
case the contract pays out in fewer states of the world. However, the price
reduction is less pronounced if the volatility parameter ζ is large. To put it
differently: an increase in ζ transfers weight to the tails of the distribution
and raises the price of the derivative Z_0 for a given positive F.

Similar phenomena to those described above can be observed in a BS world.
The price of a derivative that is linear in the asset price, for example a
forward contract, is insensitive to the volatility of the underlying asset. On
the other hand, volatility sensitivity of a plain vanilla European call option

is a positive function of the strike price, which coincides with the fact that the greater the floor F of the liquidity derivative, the higher the sensitivity of the price with respect to changes in volatility.

The parameter $\lambda^{(\rho)}$ influences the drift of the liquidity process under the risk-neutral measure. Additionally, if $\lambda^{(\rho)} \neq 0$, the volatility parameter ζ appears in the risk-neutral drift, see equation (1.21). If $\lambda^{(\rho)} > 0$, an increase in ζ causes an increase in κ^* and a reduction in θ^*. Both aspects reduce the price of the derivative Z_0. The opposite effects can be observed for $\lambda^{(\rho)} < 0$.

1.7.4   Market Price of Liquidity Risk

1.7.4.1   Overview

The recent literature provides strong evidence that investors care about liquidity risk, which implies that this type of risk is priced into asset returns. For example, Acharya and Pedersen (2003) analyze daily return and volume data from 1962 until 1999 for all common shares listed on the NYSE and the AMEX. They corroborate the hypothesis that investors require a premium for a security that is illiquid when the market as a whole is illiquid and that investors are willing to pay a premium for a security that is liquid when stock returns are low. Moreover, they show that investors are willing to pay a premium for a security with a high return when the market is illiquid. This is empirically supported by Pástor and Stambaugh (2002) who use monthly data for 34 years of common stocks traded on NYSE, AMEX, and NASDAQ. They find that stocks that are more sensitive to aggregated market liquidity have higher expected returns. A comprehensive review of theoretical and empirical approaches can be found in Pritsker (2002).

From this perspective and with the results of Subsection 1.7.2 in mind it seems important to provide a reasonable approach to determine the market price of liquidity risk from traded instruments. As an example of such a traded claim we consider a plain vanilla European put option that gives the holder the right to sell ϕ_0 units of the stock for the strike price X at maturity T. In the following we make four assumptions to simplify the explanation. Specifically, we suppose that:

- the contract is settled via physical delivery of the shares

- the day of execution coincides with the day of delivery (note that in practice there may be some exchange trading days in between)

- the short seller of the put liquidates the ϕ_0 units of the stock immediately upon exercise of the option

- neither the long nor the short party trades the underlying asset before the maturity date of the option.

The first assumption is met at many exchanges for equity options. The second and the third assumption are not critical and can be relaxed if a more general framework is desired. The last assumption ensures that for $\gamma = 0$ the underlying price process and the liquidity process are independent. Easing this restriction, for example by assuming that the short seller hedges the option in the stock market, results in more complex dynamics for the underlying asset as we have seen before when considering general trading strategies of a large investor.

1.7.4.2 Intuition

Assume that the holder of the put has to liquidate a long position in the underlying asset at a certain time T. For this purpose the investor can either sell the securities at the exchange or execute the put option and deliver the stocks to the short seller of the put. Since the market for the underlying asset is illiquid, the terminal payoff of the put is given by $\mathcal{P}_T = \max[X - S_T(1 - \rho_T\phi_0), 0]$. This can be explained as follows.

The execution of the put is optimal if it provides higher revenues than the liquidation of the underlying asset in the market. In the latter case one will receive $S_T(1 - \rho_T\phi_0)$ due to the market illiquidity. Thus, it is optimal to exercise the put option if and only if $X > S_T(1 - \rho_T\phi_0)$. If the put is exercised the short party faces a liquidity risk. Assume that the underlying asset is liquidated immediately after the execution of the put. Then the short seller has to bear a liquidity discount amounting to $S_T\rho_T\phi_0$.

The story is similar to the liquidity derivative considered in the previous section for a floor $F \equiv 0$. If the stop loss order is executed at $\tau \leq T$, the holder of the liquidity derivative would sell the underlying asset at the market for $\bar{S}(1 - \rho_\tau\phi_0)$ and receive the payoff of the derivative $\bar{S}\rho_\tau\phi_0$. In total he or she would obtain $\bar{S}\phi_0$. On the other hand, if an investor is hedged by the European put option, he or she has to deliver the shares to the short seller at T and receives $X\phi_0$.

Although the payoff of the European put option and the liquidity derivative seem closely related to each other for $X = \bar{S}$, there are still important differences. The European put option matures at time T. Therefore, it may be a suitable instrument for an investor who has to liquidate stocks at a fixed point in time. In contrast to that, the liquidity derivative matures at the stopping time $\tau \leq T$, when the stock price hits $\bar{S}$ for the first time. Considering this particular trading strategy a European put option would not provide a perfect hedge. Imagine, for example a situation where $\bar{S}$ is hit at time $\tau < T$, but then the stock price rises again and we observe $S_T > \bar{S}$. In this case the put would not be exercised at T, although the stop loss order had been executed. Only for $\tau = T$ (which is a zero probability event) would the investor be indifferent between the two claims.

One may argue that an American put option better meets the needs of an investor with a stop loss strategy than a European put, since the American put option could be exercised at any stopping time $\tau \leq T$. However, early execution of an American put option following this particular strategy does not have to be optimal. Hence, the American put option will be more expensive than the liquidity derivative. Thus, a long position in the liquidity derivative would be the cheapest way to hedge against the liquidity risk arising from a stop loss trading strategy.

1.7.4.3 Pricing a European Put on an Illiquid Stock

This subsection proposes an innovative approach for the calculation of the market price of liquidity risk. The price of a European put option can be computed analytically in our framework. The put price is not unique and depends on the market price of liquidity risk, since the market is incomplete. Therefore equating the theoretical prices to the observed market prices allows us to extract the market price of liquidity risk.

As before, the underlying asset follows a geometric Brownian motion with drift r under the risk-neutral measure. For the liquidity process we again consider a CIR process. By fixing a risk-neutral measure $\hat{P}$ the pricing formula is given by

$$
\begin{aligned}
\mathcal{P}_t &= e^{-r(T-t)} \hat{E}\left[\max[X - S_T(1 - \rho_T\phi_0), 0]\right] \\
&= e^{-r(T-t)} \hat{E}\left[(X - S_T(1 - \rho_T\phi_0))I(X > S_T(1 - \rho_T\phi_0))\right] \\
&= e^{-r(T-t)} X\hat{P}(X > S_T(1 - \rho_T\phi_0)) \\
&\quad - e^{-r(T-t)} \hat{E}\left[S_T(1 - \rho_T\phi_0)I(X > S_T(1 - \rho_T\phi_0))\right] \\
&= e^{-r(T-t)} X\hat{P}(X > S_T(1 - \rho_T\phi_0)) \\
&\quad - e^{-r(T-t)} \hat{E}\left[S_T I(X > S_T(1 - \rho_T\phi_0))\right] \\
&\quad + e^{-r(T-t)} \hat{E}\left[S_T\rho_T\phi_0 I(X > S_T(1 - \rho_T\phi_0))\right].
\end{aligned}
$$

Let f and g denote the risk-neutral density of the asset price process and of the CIR process at T, respectively. The market price of liquidity risk is contained in the (risk-neutrally) adjusted parameters of the CIR density, compare Subsection 1.7.1. In integral representation – assuming $\gamma = 0$, i.e. independence of S and ρ – we obtain

$$
\begin{aligned}
\mathcal{P}_t &= e^{-r(T-t)} X \int_{X > s(1-\rho\phi_0)} f(s)g(\rho)\mathrm{d}s\mathrm{d}\rho \\
&\quad - e^{-r(T-t)} \int_{X > s(1-\rho\phi_0)} sf(s)g(\rho)\mathrm{d}s\mathrm{d}\rho \\
&\quad + e^{-r(T-t)}\phi_0 \int_{X > s(1-\rho\phi_0)} s\rho f(s)g(\rho)\mathrm{d}s\mathrm{d}\rho \\
&= e^{-r(T-t)} X \int_0^X \left(\int_0^\infty \frac{1}{1-\rho\phi_0} f(z/(1-\rho\phi_0))g(\rho)\mathrm{d}\rho \right) \mathrm{d}z
\end{aligned}
$$

$$- e^{-r(T-t)} \int_0^X \left(\int_0^\infty \frac{z}{(1 - \rho\phi_0)^2} f(z/(1 - \rho\phi_0)) g(\rho) \mathrm{d}\rho \right) \mathrm{d}z$$

$$+ e^{-r(T-t)} \phi_0 \int_0^X \left(\int_0^\infty \frac{z\rho}{(1 - \rho\phi_0)^2} f(z/(1 - \rho\phi_0)) g(\rho) \mathrm{d}\rho \right) \mathrm{d}z,$$

using the substitution $z = s(1 - \rho\phi_0)$ in the last step. In order to determine the market price of liquidity risk one has to find the root(s) of the equation

$$\mathcal{P}_0^{\mathrm{obs}} - \mathcal{P}_0 \left(\lambda^{(\rho)} \right) \overset{!}{=} 0.$$

1.8 Conclusion

Criteria such as consistency with empirical phenomena, flexibility, and also computational aspects should be considered in the development of a liquidity model. We have presented a framework that we feel meets these requirements and has proved worthy of investigation.

This chapter introduces a continuous-time model for an illiquid market, where the trading strategy of a large investor can move prices. The innovative features of our setup include on the one hand, a time-varying market depth and on the other hand, the modeling of liquidity feedback effects.

We have analyzed two basic types of trading strategies. For positive feedback strategies and non-positively correlated Brownian motions, volatility generally is increased compared with BS. For contrarian feedback strategies one can observe basically the opposite. The picture can change completely if the asset becomes highly illiquid and the asset price dynamics are dominated by the trading activities of the large investor. These features are discussed both analytically and on the basis of simulation results.

Furthermore, an application of the general framework is proposed. We derive a closed-form expression for the price of a liquidity option, the payoff of which depends on the price difference between a stop loss limit and the average execution price. Furthermore, a pragmatic way to determine the market price of liquidity risk from traded European plain vanilla put options is presented.

Further research may focus on derivative pricing under the effective price processes or on the application of hedging strategies from incomplete markets to this setting. An investigation of optimal liquidation strategies in the proposed framework also seems promising.

1.9 Appendix

1.9.1 Proof of Proposition 1.1

Using Itô's lemma for $\phi(t, S, \rho)$ gives

$$\mathrm{d}\phi = \phi_t\mathrm{d}t + \phi_S\mathrm{d}S + \phi_\rho\mathrm{d}\rho + \phi_{S\rho}\mathrm{d}[S, \rho]_t + \frac{1}{2}\left(\phi_{SS}\mathrm{d}[S]_t + \phi_{\rho\rho}\nu^2\mathrm{d}t\right).$$

Plugging this into equation (1.4) we get

$$
\begin{aligned}
\mathrm{d}S_t &= (\sigma + \rho\phi_\rho\nu\gamma)S_t\mathrm{d}W_t + \rho\phi_\rho\nu S\mathrm{d}\bar{W}_t + \rho_t S_t\Big((\mu_t S_t + \phi_t)\mathrm{d}t + \phi_S\mathrm{d}S \\
&\quad + \phi_\rho\mathrm{d}\rho + \nu\phi_{S\rho}\mathrm{d}[S, \bar{W}]_t + \frac{1}{2}\left(\phi_{SS}\mathrm{d}[S]_t + \phi_{\rho\rho}\nu^2\mathrm{d}t\right)\Big)
\end{aligned}
$$

which in turn implies

$$
\begin{aligned}
\mathrm{d}S_t &= \frac{1}{1 - \rho S\phi_S}\Bigg(\left((\sigma + \rho\phi_\rho\nu\gamma)S_t\mathrm{d}W_t + S_t\left(\rho\phi_\rho\nu\sqrt{1 - \gamma^2}\right)\mathrm{d}\bar{W}_t\right. \\
&\quad + \rho S_t\left[\left(\frac{\mu_t}{\rho} + \phi_t + \beta(\rho_t, t)\phi_\rho + \frac{1}{2}\phi_{\rho\rho}\nu^2\right)\mathrm{d}t\right. \\
&\quad \left.\left. + \frac{1}{2}\phi_{SS}\mathrm{d}[S]_t + \nu\phi_{S\rho}\mathrm{d}[S, \bar{W}]_t\right]\right),
\end{aligned}
$$

assuming $\rho_t S_t\phi_S < 1$. Using the trial solution

$$\mathrm{d}S_t = b(t, S_t, \rho_t)\mathrm{d}t + v(t, S_t, \rho_t)S_t\mathrm{d}W_t + \bar{v}(t, S_t, \rho_t)S_t\mathrm{d}\bar{W}_t$$

and comparing coefficients leads to

$$
\begin{aligned}
v(t, S, \rho) &= \frac{\sigma + \rho\phi_\rho\nu\gamma}{1 - \rho S\phi_S} \\
\bar{v}(t, S, \rho) &= \frac{\nu\rho\phi_\rho\sqrt{1 - \gamma^2}}{1 - \rho S\phi_S}
\end{aligned}
$$

$$b(t, S, \rho)dt = \frac{\rho S}{1 - \rho S \phi_S} \left[\left(\frac{\mu_t}{\rho} + \phi_t + \beta(\rho_t, t)\phi_\rho + \frac{1}{2}\phi_{\rho\rho}\nu^2 \right) dt \right.$$

$$\left. + \frac{1}{2}\phi_{SS}d[S]_t + \nu\phi_{S\rho}\,d[S, \bar{W}]_t \right]$$

$$= \frac{\rho S}{1 - \rho S \phi_S} \left[\left(\frac{\mu_t}{\rho} + \phi_t + \beta(\rho_t, t)\phi_\rho + \frac{1}{2}\phi_{\rho\rho}\nu^2 \right) \right.$$

$$+ \frac{1}{2}\phi_{SS}S^2 \frac{\sigma^2 + \rho^2\nu^2\phi_\rho^2 + 2\gamma\rho\sigma\nu\phi_\rho}{(1 - \rho S \phi_S)^2}$$

$$\left. + \nu\phi_{S\rho}S \frac{\nu\rho\phi_\rho + \gamma\sigma}{1 - \rho S \phi_S} \right] dt,$$

since

$$d[S]_t = S^2(\nu^2 + \bar{\nu}^2)dt$$

$$= S^2 \left(\frac{\sigma^2 + \rho^2\nu^2\phi_\rho^2 + 2\gamma\rho\sigma\nu\phi_\rho}{(1 - \rho S \phi_S)^2} \right) dt,$$

and

$$d[S, \rho]_t = S\nu(\gamma\nu + \sqrt{1 - \gamma^2}\bar{\nu})dt$$

$$= S\nu \left(\frac{\nu\rho\phi_\rho + \gamma\sigma}{1 - \rho S \phi_S} \right) dt.$$

1.9.2   Functional Form of $\phi(S, \rho)$

In order to incorporate the idea that ϕ is a function of both S and ρ we choose a product approach by separating the strategy with respect to S and ρ:

$$\phi(S, \rho) = a\psi(S)\chi(\rho).$$

A function that is able to model the features described in Section 1.6 is the incomplete gamma function, defined by

$$\Gamma(x, z) = \int_z^\infty e^{-t}t^{x-1}dt.$$

We read $\Gamma(x, z)$ as a function of z for fixed x-values. It is monotonically decreasing to zero for $z \to \infty$. The incomplete gamma function captures the dependency of the trading strategy with respect to ρ for fixed S. For the contrarian feedback strategy the scenario is similar with respect to S for fixed ρ. Since the image is reversed for the positive feedback strategy the functional form has to be adjusted. Therefore we multiply the gamma function by -1 and then shift it to the positive quadrant. Thus, we choose the following representation for the dependency on ρ:

$$\chi(\rho) = \Gamma(b_1, c_1\rho),$$

and for the dependence on S:

$$\psi(S) = \begin{cases} \Gamma(b_2, c_2 S) & \text{(contrarian strategy)}, \\ \\ \Gamma(b_2) - \Gamma(b_2, c_2 S) & \text{(positive strategy)}. \end{cases}$$

Consequently, the contrarian feedback strategy is given by

$$\phi^-(S, \rho) = a\Gamma(b_1, c_1\rho)\Gamma(b_2, c_2 S), \qquad (1.24)$$

whereas the positive feedback strategy is given by

$$\phi^+(S, \rho) = a\Gamma(b_1, c_1\rho)\left[\Gamma(b_2) - \Gamma(b_2, c_2 S)\right], \qquad (1.25)$$

where $\Gamma(x)$ denotes the standard gamma function. For the simulation in Section 1.6 we use the following parameter constellation:

$$\begin{aligned} b_1 &= 2.1 & b_2 &= 6.0 \\ c_1 &= 5.5 & c_2 &= 0.05 & a &= -0.05. \end{aligned}$$

References

Acharya, V. V. and L. H. Pedersen (2003): Asset pricing with liquidity risk, Stern School of Business, New York University, USA.

Allen, F. and D. Gale (1992): Stock-price manipulation, *The Review of Financial Studies* 5, 503–29.

Almgren, R. and N. Chriss (2000): Optimal execution of portfolio transactions, *Journal of Risk* 3, 5–39.

Bank, P. and D. Baum (2002): Hedging and portfolio optimization in financial markets with a large trader, Humboldt University, Berlin, Germany.

Bertsimas, D. and A. W. Lo (1998): Optimal control of execution, *Journal of Financial Markets* 1, 1–50.

Black, F. and M. Scholes (1973): The pricing of options and corporate liabilities, *Journal of Political Economy* 81, 637–54.

Borodin, A. N. and P. Salminen (1996): *Handbook of Brownian Motion – Facts and Formulae*, Birkhäuser Verlag, Basel, Boston, Berlin.

Coughenour, J. and K. Shastri (1999): Symposium on market microstructure: a review of empirical research, *The Financial Review* 34, 1–28.

Cox, J. C., J. E. Ingersoll, and S. A. Ross (1985): A theory of the term structure of interest rates, *Econometrica* 53(2), 385–407.

Cuoco, D. and J. Cvitanić (1998): Optimal consumption choices for a large investor, *Journal of Economic Dynamics and Control* 22, 401–36.

Cvitanić, J. and J. Ma (1996): Hedging options for a large investor and forward-backward SDE's, *Annals of Applied Probability* 6, 370–98.

Dubil, R. (2002): Optimal liquidation of large security holdings in thin markets, University of Connecticut, Storrs, USA.

Ericsson, J. and O. Renault (2003): Liquidity and credit risk, McGill University, Montreal, Canada.

Frey, R. (1998): Perfect option replication for large traders, *Finance and Stochastics* 2, 115–42.

Frey, R. (2000): Market illiquidity as a source of model risk in dynamic hedging, in *Model Risk*, ed. by R. Gibson, RISK Publications, London, 125–36.

Frey, R. and P. Patie (2001): Risk management for derivatives in illiquid markets: a simulation study, RiskLab, ETH-Zentrum, Zürich, Switzerland.

Hisata, Y. and Y. Yamai (2000): Research toward the practical application of liquidity risk evaluation methods, *Monetary and Economic Studies*, December, 83–128.

Jarrow, R. (1992): Market manipulation, bubbles, corners, and short squeezes, *Journal of Financial and Quantitative Analysis* 27, 311–36.

Kampovsky, A. and S. Trautmann (2000): A large trader's impact on price processes, Johannes-Gutenberg-Universität, Mainz, Germany.

Kyle, A. (1985): Continuous auctions and insider trading, *Econometrica* 53, 1315–35.

Liu, H. and J. Yong (2004): Option pricing with an illiquid underlying asset market, Washington University, USA and University of Central Florida, Orlando, USA.

Mönch, B. (2003): Optimal liquidation strategies, Goethe University, Frankfurt am Main, Germany.

Pástor, L. and R. F. Stambaugh (2002): Liquidity risk and expected stock returns, Graduate School of Business, University of Chicago, USA.

Pritsker, M. (2002): Large investors and liquidity: A review of the literature, Federal Reserve Board, Washington, USA.

Ranaldo, A. (2000): Intraday trading activity on financial markets: The Swiss evidence, Ph.D. thesis, Université de Fribourg, Switzerland.

Schönbucher, P. J. and P. Wilmott (2000): The feedback effect of hedging in illiquid markets, *SIAM Journal of Applied Mathematics* 61(1), 232–72.

Sircar, K. R. and G. Papanicolaou (1998): General Black–Scholes models accounting for increased market volatility from hedging strategies, *Applied Mathematical Finance* 5, 45–82.

2 Optimal Liquidation Strategies

2.1 Introduction

This chapter analyzes optimal liquidation strategies for large security holdings. Order submission decisions are among the most important choices traders make. For many institutional investors (like insurance companies or pension funds) even a moderately sized position in a stock may represent a large fraction of this stock's daily trading volume. Liquidating such portfolios can incur significant costs that directly influence the return on the investment.

Nevertheless, compared with other fields of research in financial engineering, little theoretical and empirical work has been done in this area. The lack of order book data, which are necessary to calibrate appropriate models, may be one crucial reason for this fact. However, the introduction of electronic order book systems at more and more stock markets around the world and the keen competition among the large trading platforms will probably induce the exchanges in the near future to provide customers with suitable information about the instant liquidity of the listed products. The model proposed in this chapter exploits this kind of information and proposes an optimal trading strategy for a large investor.

The following sections focus on the case where a large trader has a positive initial investment in an asset and wants to close this position within a trading window of one day (the analysis for a purchasing strategy is symmetric). For this purpose he or she can submit market orders basically continuously throughout the day. Thereby the trader has to balance different issues. Obviously, the investor can liquidate the portfolio all at once immediately after the opening of the stock exchange. In this case he or she has to bear virtually no market risk, since the entries in the order book are known with certainty. If there are fixed costs on each trade, such a strategy would also minimize transaction costs of this kind. However, selling large amounts aggressively against the orders in the book will presumably have a substantial price impact and consequently lead to large liquidity discounts. Thus, the liquidation of a large portfolio is often broken up into smaller packages. Furthermore, it seems advisable to liquidate more in times when market liquidity is typically high (around noon) and less in periods where markets are usually illiquid (in the morning and before the close of the exchange). However, this procedure comes also at a cost. Forgoing the immediacy increases the market risk of the liquidation revenues. Additionally, as the number of trades increases fixed transaction costs, this component of total costs will increase as well. The model proposed in this chapter is a first approach to assess these different issues simultaneously in order to derive an optimal sales trajectory.

There is a growing theoretical literature on optimal liquidation strategies for large portfolios. Bertsimas and Lo (1998), Almgren and Chriss (2000), Hisata and Yamai (2000), and Dubil (2002) are prominent examples. The character of those papers is mainly theoretical with the intention to derive closed or semi-closed form solutions for the optimal liquidation trajectories. As mentioned above, empirical work has been hampered until recently by the lack of order book data. The main ideas of these papers are briefly outlined in Section 2.2.

This chapter focuses on a realistic modeling of intraday liquidity patterns and price impact functions for large transactions in markets, which are not perfectly liquid. Many empirical papers that investigate high-frequency order book data or transaction prices (among others see Wood, McInish, and Ord (1985), Harris (1986), Jain and Joh (1988), McInish and Wood (1990),

McInish and Wood (1992), Kirchner and Schlag (1998), and Ranaldo (2000)) find that intraday time series of volatility, trading volume, order flow, and transaction costs follow a U-shape pattern, i.e. these variables start at a high value at the opening of the trading day, fall to lower levels over the day and then rise again towards the close of trading.

The possible reasons for these intraday patterns are discussed controversially in the literature, and it seems they can be explained at least partly by the information flow through trades, resulting in progressively smaller adverse selection costs as the day evolves (see Wei (1992), Hasbrouck (1991), Foster and Viswanathan (1993), Lin, Sanger, and Booth (1995)). The decrease in liquidity before the overnight trading halt may reflect the cost of holding inventory over the upcoming nontrading period (see Amihud and Mendelson (1982), Bessembinder (1994), Lyons (1995), and Huang and Masulis (1999)). This chapter does not aim to provide yet another explanation, but rather takes these patterns as exogenous.

A recent study conducted by Linnainmaa (2003) analyzes order book data for the 30 largest stocks traded at the Helsinki Exchanges. The author provides empirical evidence that institutional investors often demand liquidity and employ market orders more intensively than less sophisticated retail investors who may be classified as net suppliers of liquidity.

Modeling and hedging aspects introduced by illiquidity and the presence of one or more large traders are discussed by Cvitanić and Ma (1996), Cuoco and Cvitanić (1998), Sircar and Papanicolaou (1998), Frey (1998, 2000), Schönbucher and Wilmott (2000), Kampovsky and Trautmann (2000), Frey and Patie (2001), Liu and Yong (2004), Bank and Baum (2002), and Esser and Mönch (2002).

The study provided in this chapter is based on an approach similar to Frey's (2000) model with constant liquidity, but introduces a time-dependent factor to model the U-shape pattern.

The rest of the chapter is organized as follows: Section 2.2 summarizes the key ideas of related papers. The dataset used throughout this chapter is introduced in Section 2.3. The general framework is presented in Section 2.4. In Section 2.5 the modeling of the impact function is discussed. The proposed setup to determine the optimal sales trajectory is calibrated in Section 2.6 for one example to give an impression of the magnitude of the liquidity discounts and the liquidation intervals. The chapter concludes in Section 2.7 with a brief summary and a discussion of issues for further research.

2.2   Related Literature

Since research conducted by Bertsimas and Lo (1998), Almgren and Chriss (2000), Hisata and Yamai (2000), and Dubil (2002) is closely related to this chapter, the results of these studies are summarized briefly in this section.

Bertsimas and Lo (1998), as well as Almgren and Chriss (2000), consider an optimization problem of an investor who has to liquidate a large position in a security within an exogenously given period of time, whereas Dubil (2002) solves for the optimal length of the period over which an investor should close a certain position in a stock.

Bertsimas and Lo (1998) derive dynamic trading strategies using stochastic dynamic programming techniques that minimize the expected cost of purchasing a large block of shares. The authors do not consider the volatility of costs for different trading strategies. They show that naive strategies of selling a constant number of shares in equally spaced time intervals are optimal only if the price impact is linear and permanent and if the stock prices follow an arithmetic random walk. The authors also consider alternative formulations of the stock price dynamics and the impact function and show that optimal liquidation strategies vary through time and may depend on market conditions and the number of shares that remain to be liquidated.

Almgren and Chriss (2000) incorporate the market risk associated with the liquidation of a position over a longer period by using a mean-variance approach. The authors construct the efficient frontier in the space of time-dependent liquidation strategies. They derive tractable analytical results for static strategies assuming an arithmetic random walk for the stock price and linear impact functions. Like Dubil (2002), they distinguish between a permanent and a temporary market impact, a concept that is discussed, for example in Holthausen, Leftwich, and Mayers (1987). The permanent price effect changes the equilibrium price due to the transaction of the large trader. The temporary impact is caused by an exhaustion of the liquidity supply when the market order of the large investor is executed. It does not influence the stock price in the next liquidation interval.

Hisata and Yamai (2000) and Dubil (2002) extend the framework of Almgren and Chriss (2000) in order to endogenize the final liquidation horizon. Both approaches consider the risk of a liquidation strategy in a Value-at-Risk (VaR) setup and assume a constant speed of trading. The two frameworks basically differ in the way they specify the dynamics of the asset price process and in the parameterization of the impact function, yet they derive similar results. In addition to the standard case, where the market impact is linear in trading volume, Hisata and Yamai (2000) provide closed-form solutions if the market impact is defined by a square root function. They also consider a stochastic market impact model and a portfolio model where the solution for the optimal liquidation horizon can be obtained numerically. Dubil (2002) assumes two parameterizations for the impact function: general power functions uncorrelated with the share price, and linear functions correlated with the price process. The first choice of the impact functions allows for the derivation of closed-form solutions for certain specialized parameterizations, while the second formulation can model a feedback loop between price dislocations and liquidity.

The general setting of this chapter is similar to the approaches of Hisata and Yamai (2000) and Dubil (2002) as the overall time needed to complete the liquidation process is determined endogenously. However, the new model allows us additionally to specify an upper bound for the final liquidation horizon. The new framework forgoes the assumption of a constant speed of trading. This enables the investor to adapt the liquidation behavior to varying levels of liquidity throughout the trading day. Furthermore, the new model considers explicitly the resiliency of the order book.

2.3   Description of the Market and Dataset

Important properties of the liquidation model proposed in this chapter will be illustrated by using real-world order book data from the German automated trading system XETRA.

XETRA is an order-driven market where investors, by placing limit orders, establish prices at which other participants can buy or sell shares. A trade takes place whenever a counterpart order hits the quotes. The system was introduced in 1997 by the German Stock Exchange. At the time the data were collected, XETRA was open for trading from 9:00 a.m. to 8:00 p.m. The trading day starts with an opening auction, followed by continuous trading, which can be interrupted by one or several intraday auction(s). At the end of the day there is either a closing or an end-of-day auction. On XETRA, all of approximately 6,000 equities listed on the Frankfurt Stock Exchange are tradable. The minimum trading volume is one share. Market participants can see all non-hidden entries on each side of the order book, but trading in XETRA is anonymous, i.e. market participants cannot identify the counterparts. On XETRA there are no dedicated providers of liquidity for blue chip stocks. For small and mid cap stocks, designated sponsors (banks and security firms) are given incentives to provide sufficient liquidity by responding to a quote request within a fixed period of time. Floor trading with market makers on the Frankfurt Stock Exchange still takes place but loses more and more market share. In the blue chip segment merely every tenth share is still traded on the floor.

Based on event histories for 61 trading days (January 03 – March 28, 2002), which were provided by the Trading Surveillance Office of the Deutsche Börse AG, order book sequences were reconstructed. Therefore, by starting from an initial state, each change in the order book depth caused by entry, filling, cancelation or expiration of orders was considered as prescribed by the market model of XETRA. This allows us to estimate the price impact function of the liquidation model. Due to the huge amount of data only the blue chip share MAN is considered as a representative example of the stocks that are traded in XETRA. The MAN corporation is one of Europe´s leading suppliers of capital goods and systems in the fields of commercial vehicle construction, and mechanical and plant engineering. Over the sample period the daily turnover in MAN on XETRA ranged from 500,000 to 1,000,000 shares, and the order book depth from 80,000 to 150,000 shares.

2.4 The Basic Setup

This section introduces the basic model and provides the motivation for the key assumptions.

The large trader holds ϕ_0 units of a security, which have to be liquidated before time T^*. The variable T denotes the actual time needed to sell the initial position ϕ_0 completely if the investor follows a certain strategy. Consequently a strategy is admissible if $T \leq T^*$. The period from time $t_0 = 0$ to time T is divided into n intervals $\tau_i = t_{i+1} - t_i$, whereas $i = 0, \ldots, n-1$, so that

$$T = t_n = \sum_{i=0}^{n-1} \tau_i.$$

In contrast to Almgren and Chriss (2000) and Dubil (2002), the intervals can have different length, but at the end of each interval the large trader liquidates some positive amount of shares $\Delta\phi_{t_i} > 0$, so that

$$\Delta\phi_{t_i} \leq \phi_0 - \sum_{j=1}^{i-1} \Delta\phi_{t_j} \qquad i = 1, \ldots, n$$

and

$$\sum_{i=1}^{n} \Delta\phi_{t_i} = \phi_0.$$

To obtain comparable flexibility one could alternatively assume equally spaced time intervals (as is done, for example in Almgren and Chriss (2000)) and allow the investor not to trade in some intervals. In this case, in order to model a representative set of liquidation trajectories attainable in practice, the length of the time intervals has to be sufficiently small. However, for $\tau \to 0$ such an assumption blows up the optimization problem, since for every time interval one has to determine the optimal quantity the investor should trade.

Similarly to Frey (2000), it is supposed that the dynamics of the asset price process S_t follow the stochastic differential equation:

$$\mathrm{d}S_t = \sigma S_{t-}\mathrm{d}W_t - S_{t-}\mathrm{d}\Psi_t^+, \tag{2.1}$$

where $S_{t-} = \lim_{s \nearrow t} S_t$ and $\mathrm{d}\Psi_t^+$ denotes the price impact of the investor's trading strategy. In this setup the stock holdings of the large investor vary discretely. Therefore, the jumps caused by the trading activity will be denoted by $\Delta \Psi_t^+ := \Psi_t^+ - \Psi_t$.

To model the price impact $\Delta \Psi_t$, a product approach is used:

$$\Delta \Psi_t = \rho \left(\Delta \phi_t \right) \vartheta \left(\tilde{t} \right),$$

so that the price impact depends both on the trade size $\Delta \phi_t$ and on the time of day $\tilde{t}$, which is measured in minutes since the opening of the exchange.

The stock price immediately before and after the jump i can be written as

$$S_{t_i} = \left[1 - \rho \left(\Delta \phi_{t_i} \right) \vartheta \left(t_i \right) \right] S_{t_i-} = \gamma_i S_{t_i-}, \tag{2.2}$$

where $\gamma_i := 1 - \rho \left(\Delta \phi_{t_i} \right) \vartheta \left(t_i \right)$.

The stock price after the trade must not be confused with the average price per unit $S_{t_i}^*$ realized by the large investor who liquidates $\Delta \phi_{t_i}$ stocks at time t_i. The average price per unit $S_{t_i}^*$ is bounded by

$$S_{t_i-} \geq S_{t_i}^* \geq S_{t_i}$$

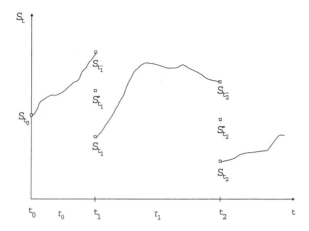

Fig. 2.1: Stock prices immediately before and after a jump due to liquidation by
the large trader.

and can be calculated as

$$S_{t_i}^* = \frac{1}{\Delta\phi_{t_i}} \int_0^{\Delta\phi_{t_i}} (1 - \rho(u)\,\vartheta(t_i))\, S_{t_i^-}\, du$$

$$= \left[1 - \frac{\vartheta(t_i)}{\Delta\phi_{t_i}} \int_0^{\Delta\phi_{t_i}} \rho(u)\, du \right] S_{t_i^-}$$

$$= \frac{1 - \frac{\vartheta(t_i)}{\Delta\phi_{t_i}} \int_0^{\Delta\phi_{t_i}} \rho(u)\, du}{1 - \rho(\Delta\phi_{t_i})\,\vartheta(t_i)}\, S_{t_i} = \delta_i\, S_{t_i}, \tag{2.3}$$

where $\delta_i := \dfrac{1 - \frac{\vartheta(t_i)}{\Delta\phi_{t_i}} \int_0^{\Delta\phi_{t_i}} \rho(u)du}{1 - \rho(\Delta\phi_{t_i})\vartheta(t_i)}$. Figure 2.1 shows this setup.

Furthermore, suppose that there are also fixed transaction costs of k per
trade. While variable transaction costs could easily be introduced by an
additional multiplicative factor in equation (2.3) and would not significantly
influence the choice of the optimal trading trajectory, fixed transaction costs

certainly do. For example, assume that the objective function of the large investor considers only the expected net liquidation value, $\mathbf{E}G$, given by

$$\mathbf{E}G = -nk + \sum_{i=1}^{n} \Delta\phi_{t_i}\,\delta_i\,\mathbf{E}S_{t_i}. \tag{2.4}$$

For $k = 0$ and a sensible definition of ρ, i.e. for $\partial\rho/\partial\phi > 0$ and $\partial^2\rho/\partial\phi^2 > 0$, one gets an optimal sales trajectory with $n \to \infty$ and $\Delta\phi_i \to 0$. If $k > 0$, the investor has to balance, on the one hand, the advantage of trading small amounts that cause only a small price impact and, on the other hand, the overall fixed transaction costs, increasing in the number of trading intervals. At this point the question may arise, which costs can be assumed to represent fixed transaction costs? Clearly, fixed brokerage fees and any fixed charges incurred by the exchange belong to this category. However, more important in daily life are the opportunity costs for handling the transactions in the front and back offices. These costs do not depend on trading volume, but predominantly on the number of transactions.

For the sake of simplicity, the risk–reward trade-off is modeled via the expected net liquidation value $\mathbf{E}G$ and the respective standard deviation $\mathrm{STD}\,(G)$. The large trader then maximizes the objective function with α as the degree of risk aversion

$$\max\left[\mathbf{E}G - \alpha\mathrm{STD}\,(G)\right] \tag{2.5}$$

by choosing the optimal sales trajectory $\{\tau_i, \Delta\phi_{t_i}, n\}^*$.

Proposition 2.1 (Expectation and standard deviation of the net liquidation value) *Suppose the stock prices satisfy the stochastic differential equation (2.1). Then the expected value and the standard deviation of the net liquidation value G are given by*

$$\mathbf{E}G = -nk + S_{t_0}\sum_{i=1}^{n} \Delta\phi_{t_i}\delta_i \prod_{j=1}^{i} \gamma_j \tag{2.6}$$

and

$$\text{var}(G) = \sum_{i=1}^{n} (\Delta\phi_{t_i}\, \delta_i)^2 \,\text{var}(S_{t_i}) \tag{2.7}$$

$$+ 2 \sum_{i<j}^{n} \Delta\phi_{t_i}\, \delta_i\, \Delta\phi_{t_j}\, \delta_j \,\text{cov}\left(S_{t_i}, S_{t_j}\right),$$

where

$$\text{cov}(S_{t_i} S_{t_j}) = S_{t_0}^2 \left[\exp\left(\sigma^2(t_i - t_0)\right) - 1\right] \prod_{m=1}^{i} \gamma_m^2 \prod_{l=i+1}^{j} \gamma_l.$$

The proof is given in Appendix 2.8.

In the proposed model, G is not exactly normally distributed, especially for small n. However, to keep the problem handy for demonstration, the higher moments of the distribution will not be considered in the objective function (2.5). A moderate number of liquidation intervals is sufficient to justify the approximation of the distribution of G by a Gaussian. This will be demonstrated by a simulation study in Section 2.6.3.

2.5 Modeling the Impact Function

The framework proposed in this chapter incorporates distinctive features in order to model stock market liquidity in a sensible way. Specifically, the model contains:

- a U-shape for intraday stock market liquidity,

- power price impact functions,

- periods that allow the order book to be rebuilt, as boundary conditions for the time intervals between subsequent trades.

The relevance of these issues will be discussed briefly in this section.

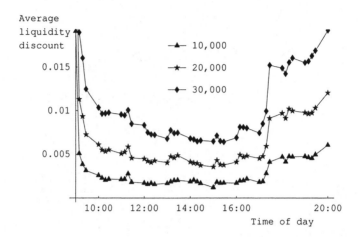

Fig. 2.2: Relative liquidity discounts for sell market orders (averages over 61 trading days) as a function of the time of day. Quantities: 10,000 (20,000,
30,000) MAN shares.

2.5.1   U-shape of Stock Market Liquidity

In the dataset used for the implementation of the proposed model one can
clearly identify a U-shape pattern in the relative liquidity discount Ψ over the
trading day (see Figure 2.2). This phenomenon is typical for stocks traded
on XETRA. The relative liquidity discount is the difference between the best
bid price and the average execution price per unit of a hypothetical market
order, divided by the best bid price.

From the opening at 9:00 a.m. to 11:00 a.m. there is a sharp decline in
the liquidity discount. From 11:00 a.m. to 5:00 p.m. it moves in a narrow
range at a low level. A significant reaction of the liquidity discount to the
stock market opening at 3:30 p.m. in New York cannot be identified. From
5:00 p.m. the average liquidity discount rises fast and remains at a high level
from 6:00 p.m. to 8:00 p.m., when the stock exchange closes.

Why is it important to consider the U-shape pattern in an intraday liquidation model? Obviously, if the investor is not too impatient, he or she should
liquidate more in periods of high liquidity (around noon) than in situations

of low liquidity (immediately after the opening and before the closing of the exchange).

A functional form that is able to model this U-shape pattern is given by $\vartheta\left(\tilde{t}\right)$ with

$$\vartheta\left(\tilde{t}\right) = \frac{d}{\tilde{t}+e} + \frac{f}{g-\tilde{t}} \, . \tag{2.8}$$

Here $\tilde{t}$ measures the time in minutes elapsed after the opening of the exchange and d, e, f, and g are positive parameters with $e < \tilde{t}^{max} < g$. The variable $\tilde{t}^{max}$ denotes the time in minutes between the opening and the close of the exchange.

2.5.2 Power Price Impact Function

The dotted line in Figure 2.3 shows the relative price impact (averaged over 61 trading days) at noon as a function of the trading volume for the dataset under consideration. It is apparent that a linear approximation (dashed line) of this function provides only a poor fit.

Exponential and power functions seem more suitable to model this relationship. For numerical convenience a simple power function is employed in the following:

$$\rho\left(\Delta\phi_t\right) = a\,\Delta\phi_t^b, \qquad a \geq 0. \tag{2.9}$$

As one can see in Figure 2.3, this function fits the data very well.

2.5.3 The Resiliency of the Order Book

The introduction of a deterministic relationship between the price impact ρ and the trade size $\Delta\phi$ implies that between two transactions the large trader has to allow the market to rebuild the order book completely. Ignoring this boundary condition in the optimization problem can result in misleading conclusions. Imagine a situation where the completion of a sell market order

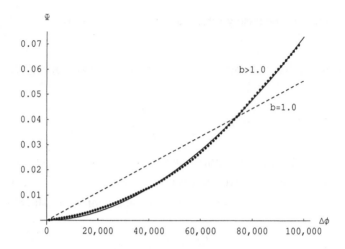

Fig. 2.3: Relative liquidity discounts at noon (averages over 61 trading days) as a function of the number of MAN shares to be liquidated. Dotted line: raw data. Dashed line: linear fit. Solid line: fitted power function.

by the large investor completely clears the order book on the bid side. Then the large trader has to wait until the bid side of the order book has recovered before he or she can sell another package of shares. Also in less extreme situations, if the market order of the large investor absorbs the order book depth only partly, being too impatient between subsequent transactions will result in deteriorating path-dependent liquidity discounts.

To operationalize this idea, two assumptions are made in the following. The order book is considered completely rebuilt after a transaction by the large investor as soon as:

- the bid–ask spread has again narrowed to the level immediately before the last transaction of the large investor, and

- the same volume that was traded by the large investor can be traded with the same or a smaller liquidity discount compared with the one the large trader had to bear in the previous transaction.

Instead of using the resiliency of the order book as a boundary condition for the length of the trading intervals, one could also separate the price im-

pact $\rho(\Delta\phi_t)$ into a permanent and a temporary component, as proposed by
Almgren and Chriss (2000), Hisata and Yamai (2000), and Dubil (2002).
However, if market orders are observed very frequently in the dataset available to calibrate the model, the permanent and the temporary component
may be difficult to separate, especially if the temporary price impacts of
different market orders overlap.

2.6   Numerical Results

The model described above was implemented using the MAN dataset for
61 trading days. We focus our analysis on continuous trading and do not
consider intraday auctions. The results are presented below.

2.6.1   Parameter Specification

Assume that $\tilde{t}$ in equation (2.8) measures the time in minutes elapsed after
9:00 a.m. The parameter values for the functions $\vartheta(\tilde{t})$ and $\rho(\Delta\phi_t)$, obtained
via least-squares estimation, are shown in Table 2.1.

The proposed functional form is able to model the empirical data quite well
$\left(R^2_{adj} = 0.964\right)$. Figure 2.4 shows the surface of the fitted function.

In order to estimate the minimum time that has to elapse between subsequent trades, all large market sell orders (order volume $\geq 1,000$ shares) in
the sample were grouped into seven intervals, assuming that the time the
market needs to recover is independent of the time of day. Table 2.2 and
Figure 2.5 show the average $\tau^{\min}$ (measured in minutes) for every interval.
Although the relationship between $\Delta\phi$ and $\tau^{\min}$ seems to be slightly concave
empirically, $\tau_i^{\min}$ is modeled as a linear function of $\Delta\phi_i$ in order to avoid
nonlinear constraints in the optimization problem:

$$\tau_i^{\min} = \beta\,\Delta\phi_i. \qquad (2.10)$$

The estimation of the parameter β in equation (2.10) via least squares yields
an estimate of $\hat{\beta} = 0.00098$ and an R^2 of 0.83. If accuracy is crucial and

Table 2.1: Estimated parameters for the functions $\vartheta\left(\tilde{t}\right)$ and $\rho\left(\Delta\phi_t\right)$.

	Estimate	Asymp. SE	t-statistics
a	0.42322	$1.34534 *10^{-12}$	$3.14583 *10^{11}$
b	1.45806	0.00911	159.995
d	$4.22391 *10^{-7}$	$4.24458 *10^{-8}$	9.95129
e	17.68900	0.29029	60.93590
f	$2.34585 *10^{-6}$	$2.35274 *10^{-7}$	9.97072
g	837.27800	3.13531	267.04800

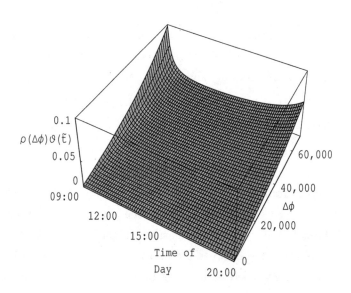

Fig. 2.4: The fitted surface $\rho\left(\Delta\phi_t\right)\vartheta\left(\tilde{t}\right)$.

Table 2.2: Average time $\tau^{\min}$ in minutes required for recovery of the order book after the execution of a large market order.

	Midpoint of interval	$\tau^{\min}$
$1{,}000 \leq \Delta\phi < 2{,}000$	1,500	2.3263
$2{,}000 \leq \Delta\phi < 3{,}000$	2,500	3.0703
$3{,}000 \leq \Delta\phi < 4{,}000$	3,500	4.3802
$4{,}000 \leq \Delta\phi < 5{,}000$	4,500	5.3110
$5{,}000 \leq \Delta\phi < 10{,}000$	7,500	11.0917
$10{,}000 \leq \Delta\phi < 15{,}000$	12,500	14.5073
$15{,}000 \leq \Delta\phi < 25{,}000$	20,000	16.1208

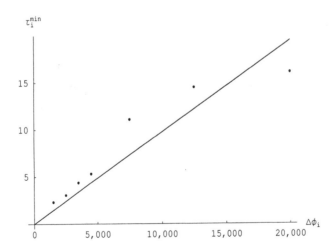

Fig. 2.5: The resiliency of the order book. Fitted function to model the time in minutes to be elapsed between subsequent trades.

computational time is cheap, one should define $\tau^{\min}$ as a nonlinear function, which may provide a higher R^2.

For the estimation of the volatility parameter in the stock price dynamics given in equation (2.1) best bid quotes, collected at intervals of 15 minutes from 9:00 a.m. to 8:00 p.m., were used and 15-minute returns for the best bid price were computed. Overnight price changes were not taken into consideration. The estimation for the volatility parameter yields $\sigma = 0.7$. Furthermore, S_0 is set to € 28.55 which is the closing price on March 28, 2002, the last day of the sample period.

2.6.2 Numerical Examples

For the first example, suppose that the large trader does not care about the risk of the liquidation revenues ($\alpha = 0$). Consequently, the investor will choose the sales trajectory that maximizes the expected liquidation value. Furthermore, assume that the investor has to liquidate 50,000 MAN shares within one trading day (approximately 5–10% of daily turnover) and that each transaction incurs fixed costs of € 50.00. To solve the optimization problem, the sequential quadratic programming algorithm e04ucc of the NAG C Library was used. It is designed to minimize an arbitrary smooth function subject to simple bounds on the variables, linear constraints, and smooth nonlinear constraints. To find the optimal sales trajectory, the following problem was solved for each $n = 1, \ldots, n^{\max}$:

$$\max_{\{\tau_i, \Delta\phi_i\}} \quad \mathbb{E}G$$

$$
\begin{aligned}
\text{s.t.} \quad \Delta\phi_{t_i} &> 0 \\
\sum_{i=1}^{n} \Delta\phi_{t_i} &= \phi_0 \\
\tau_i &\geq \beta\Delta\phi_{t_i} \\
\tau_0 &> 0 \\
\sum_{i=0}^{n-1} \tau_i &\leq T^*.
\end{aligned}
$$

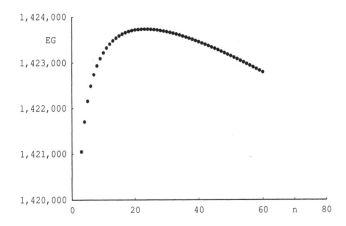

Fig. 2.6: Expected liquidation revenues as a function of n for optimal τ_i and $\Delta\phi_{t_i}$.

Then, the optimal n^* that provides the highest value of the objective function was identified. Figure 2.6 illustrates the result. The optimal sales trajectory comprises 23 intervals and provides an expected liquidation value of € 1,423,734.72. Reducing or increasing the number of intervals decreases the expected liquidation revenues, for example by € 2,686.81 or 0.2% if the portfolio is otherwise optimally liquidated in three steps. Selling the whole amount of 50,000 shares all at once at the opening of the exchange reduces the expected liquidation revenues by € 15,944.72 or 1.1% to € 1,407,740.00.

Figure 2.7 summarizes the optimal strategy. The investor should start the liquidation process at 12:40 p.m. and then increase the packages that are sold step by step as market liquidity is improving. The linear constraint (2.10) controlling the time needed to rebuild the order book is binding in almost all cases, so that the length of the liquidation intervals is increasing as well. At this point the question may arise of why the amounts the large investor should liquidate in each interval are strictly increasing. If the large trader considers only the expected liquidation value and if one observes an intraday U-shape pattern in liquidity discounts, one may expect an inverse U-shape pattern in the liquidation amounts. However, each transaction of the large

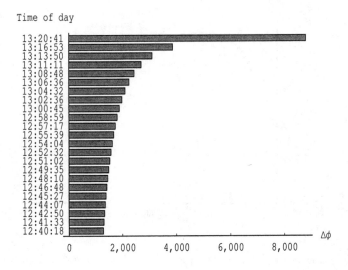

Fig. 2.7: Fitted function to model the time in minutes that should elapse between subsequent trades.

trader has a negative price impact that adversely influences the liquidation revenues of the following liquidation intervals. Taking this into consideration, the large trader should liquidate a substantial part of the overall position in the last interval, since the resulting negative price impact of this transaction no longer affects his or her liquidation revenues. In the example this seems to be the dominant effect also, since the slope of the intraday liquidity curve is flat around noon.

To compare this example with other parameterizations, the duration D of the liquidation process is calculated. It is defined as

$$
\begin{aligned}
D &= \tau_1 \frac{\Delta\phi_{t_1}}{\phi_0} + (\tau_1 + \tau_2) \frac{\Delta\phi_{t_2}}{\phi_0} + \ldots + (\tau_1 + \tau_2 + \ldots + \tau_n) \frac{\Delta\phi_{t_n}}{\phi_0} \\
&= \sum_{j=1}^{n} \left(\sum_{i=1}^{j} \tau_i \right) \frac{\Delta\phi_{t_j}}{\phi_0}.
\end{aligned}
\tag{2.11}
$$

It identifies the average weighted time needed to liquidate the position and must not be confused with the overall time needed to complete the liquidation process. The latter measure does not weight the intervals with the respective

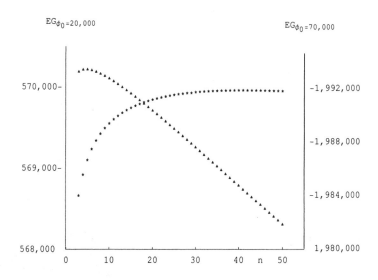

Fig. 2.8: Expected liquidation revenues for $\phi_0 = 20,000$ (left scale) and $\phi_0 = 70,000$ (right scale) shares as a function of n, if τ_i and $\Delta\phi_{t_i}$ are optimally chosen.

liquidation volume, it just sums them up. In the example one obtains $D = 4h$, 3min, 12sec. Thus, if the large trader does not care about the risk of the liquidation value ($\alpha = 0$), he or she should sell the portfolio around noon, when market liquidity is comparatively high. Later on, one can see that the durations of the optimal liquidation strategies decline as the investor becomes more risk averse.

In the next example the initial stock holdings ϕ_0 are varied. Figure 2.8 shows the results. As a rule of thumb one can state that the larger ϕ_0, the more often the investor should trade. In the case where $\phi_0 = 20,000$ the portfolio should be liquidated in five packages. If $\phi_0 = 70,000$, the investor should trade 41 times in order to realize the highest expected liquidation value. Compared with the first example, the duration of the liquidation process changes only slightly to $D = 3h$, 58min, 49sec for $\phi_0 = 20,000$ and to $D = 4h$, 6min, 16sec for $\phi_0 = 70,000$.

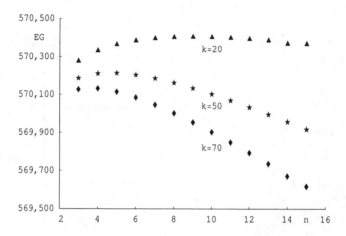

Fig. 2.9: Expected liquidation revenues for $k = 20$, $k = 50$, and $k = 70$ as a function of n, if τ_i and $\Delta\phi_{t_i}$ are optimally chosen ($\phi_0 = 20{,}000$).

The amount of fixed transaction costs influences the optimal number of trading intervals as well. Figure 2.9 provides an insight into this issue. Assume for this example that the initial stock holdings are $\phi_0 = 20{,}000$. If fixed transaction costs are increasing, the optimal number of intervals decreases. For $k = 20$, the optimal strategy consists of 9 intervals, for $k = 50$, of 5 intervals, and for $k = 70$, of only 4 intervals. Again, the duration of the liquidation process is quite insensitive to a change of fixed transaction costs. For $k = 20$, $k = 50$, and $k = 70$, one obtains $D = 3$h, 59min, 07sec, $D = 3$h, 58min, 49sec, and $D = 3$h, 58min, 40sec respectively.

In the last example the parameter α is varied. Figure 2.10 corroborates the initial guess that the portfolio should be liquidated in smaller packages if the aversion towards risk increases. An investor who trades more often reduces the price risk of the liquidation process, since the probability of observing a low stock price at a single point in time is higher than the joint probability of observing low stock prices several times a day. For $\alpha = 0.0$, 0.25, 0.5, and 1.0 the optimal number of trading intervals is given by 23, 35, 37, and 48, respectively. The impact of the parameter α on the liquidation strategy becomes apparent in more detail if the duration of the liquidation process is investi-

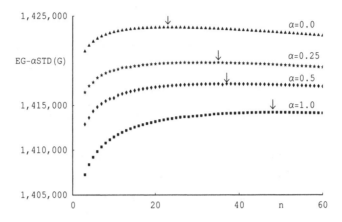

Fig. 2.10: Values of the objective function for $\alpha = 0.0$, 0.25, 0.5, and 1.0 as a function of n, if τ_i and $\Delta\phi_{t_i}$ are chosen optimally ($\phi_0 = 50,000$). The arrows point to the optimal number of liquidation intervals for a given value of α.

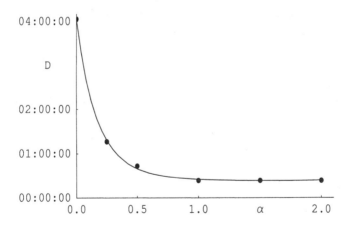

Fig. 2.11: Duration of the liquidation process as a function of α, if τ_i and $\Delta\phi_{t_i}$ are chosen optimally ($\phi_0 = 50,000$).

gated. Figure 2.11 illustrates this result. The more α increases, the more impatient the investor becomes and the earlier the portfolio should be sold. Two cost components reduce the expected liquidation value in this case. First, fixed transaction costs are rising, since the investor trades more often. Second, the trader has to accept higher liquidity discounts due to the U-shape of intraday market liquidity.

For the examples presented above it was assumed that the large investor solves for the optimal trading strategy at the beginning of the trading day. However, this assumption is not critical and can easily be relaxed by adjusting the parameters e and g in equation (2.8) if the decision on how to liquidate the portfolio has to be made at another point in time.

2.6.3 The Distribution of Liquidation Revenues

In order to get an impression of the distribution of the liquidation value and to justify the implicit assumption of normally distributed liquidation revenues, 1,000,000 outcomes of the liquidation value G are simulated. For this purpose the stochastic process given in equation (2.1) is discretized using an Euler scheme. The parameter α in the objective function is set to 1.64 and it is assumed that the large investor follows the optimal trading strategy to liquidate $\phi_0 = 50{,}000$ MAN shares with $k = 50$. Thus, the value of the objective function can be interpreted as the lower 5% quantile of the liquidation value, if G was normally distributed. For the given choice of parameters the optimization yields $n^* = 47$, $\mathbf{E}G = 1{,}419{,}672.98$, and $\mathrm{STD}(G) = 5{,}458.01$.

Figure 2.12 shows the results. For the simulated liquidation value G the gray line shows the approximated density obtained by applying standard kernel estimation procedures. The black dots represent the estimated Gaussian density function for $\mu = 1{,}419{,}672.98$ and $\sigma = 5{,}458.01$. Obviously, the differences between the two distributions are negligible. The hypothesis that the two distributions are different is tested via the Kolmogorov–Smirnov test, and the null cannot be rejected at the 5% level. Comparing the lower 5% quantile of the simulated distribution 1,410,729.35 with the optimal value of

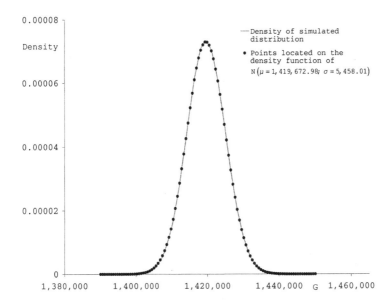

Fig. 2.12: The gray line shows the approximated density of the simulated liquidation value for $\phi_0 = 50{,}000$, $k = 50$, and $\alpha = 1.64$. The black dots represent some values of the Gaussian density function for $\mu = 1{,}419{,}672.98$
and $\sigma = 5{,}458.01$, the parameter values that were obtained for the
optimal liquidation strategy.

the objective function 1,410,695.36 one can state that the downside risk of
the strategy is slightly overestimated if the value of the objective function
is interpreted as a value of risk. However, for most practical purposes this
difference is negligible and the model seems well suited for VaR calculations.

2.7 Conclusion and Issues for Further Research

This chapter operationalizes the concept of intraday market liquidity patterns in the determination of optimal liquidation strategies for large security
holdings. Innovative features of this setup include the explicit modeling of
an intraday U-shape pattern, the consideration of the resiliency of the order book through a time constraint in the objective function of the large

investor, and the introduction of a permanent price impact function that fits empirical order book data very well. The proposed model is far more flexible than other approaches suggested in the literature, as it allows for liquidation intervals of varying length and relaxes the assumption of a constant speed of trading. Due to the formulation of the objective function the model is especially suited for VaR-regulated institutions, which want to extend the short-term management of market risk by an exogenous liquidity factor. The implementation of the model with real-world order book data shows that the model is computationally tractable and able to provide intuitively reasonable results.

A sensitivity analysis shows that if the number of initial stock holdings is increased, the optimal number of liquidation intervals will also rise. If fixed transaction costs are reduced, the large investor obtains more flexibility and the optimal number of liquidation intervals increases. By varying the agent's preference towards risk it becomes apparent that the investor can use the flexibility to limit the risk of the liquidation revenues if he or she trades more often. Additionally, if the trader becomes more risk averse, increasing liquidity discounts due to an earlier liquidation of the portfolio have to be balanced against the risk of the liquidation value. One can observe a decreasing duration of the liquidation process in this situation.

In contrast to the papers by Almgren and Chriss (2000), Hisata and Yamai (2000), and Dubil (2002) the price impact is not separated into a permanent and a temporary component. Instead it is assumed that the whole price impact is permanent and influences the liquidation revenues in the following liquidation intervals. The reason for this rather strong assumption is the lack of an appropriate procedure to separate the permanent and the temporary impact empirically from electronic order book data. Existing approaches (e.g. the one suggested by Holthausen, Leftwich, and Mayers (1987)) that split the price impact into a permanent and a temporary component assume that the temporary price impact lasts for a certain period of time, for example until the end of the trading day. This definition may be acceptable for very large transactions. However, for medium-sized trades that occur

more frequently, further research toward a comprehensive method has to be
conducted, so that one can separate overlapping temporary price impacts
resulting from different transactions.

When applying the proposed framework one should be aware that the only
source of randomness modeled in this setup is the stock price. Both the
market depth and the time needed for recovery of the order book are assumed
to be deterministic, while in practice one may recognize significant variations
of these variables through time. Further research should therefore focus on
extensions of the approach presented in this chapter, e.g. along the lines of
the stochastic liquidity model proposed by Esser and Mönch (2002).

2.8   Appendix: Derivation of $\mathbb{E}G$ and $\mathrm{var}\,(G)$

A simple way to compute the expectation is to use an iterative decomposition
of the path of S, given the new starting point after a jump downwards. Using
(2.2) one can write S_{t_i} as

$$
\begin{aligned}
S_{t_i} &= \gamma_i\, S_{t_i-} \\
&= \gamma_i\, S_{t_{i-1}} \exp\left[-1/2\sigma^2(t_i - t_{i-1}) + \sigma(W_{t_i} - W_{t_{i-1}})\right]
\end{aligned}
$$

since S_{t_i-} is log-normally distributed. Iterating this procedure yields

$$
\begin{aligned}
S_{t_i} &= \gamma_1 \ldots \gamma_i\, S_{t_0} \exp\left[-1/2\sigma^2[(t_i - t_{i-1}) + \ldots + (t_1 - t_0)]\right] \\
&\quad \exp\left[\sigma(W_{t_i} - W_{t_{i-1}}) + \ldots + \sigma(W_{t_1} - W_{t_0})\right] \\
&= \gamma_1 \ldots \gamma_i\, S_{t_0} \exp\left[-1/2\sigma^2(t_i - t_0)\right] \exp\left[\sigma(W_{t_i} - W_{t_{i-1}})\right] \\
&\quad \ldots \exp\left[\sigma(W_{t_1} - W_{t_0})\right].
\end{aligned} \tag{2.12}
$$

This yields the following expectation value using the independence
of the increments of the Brownian motion and the fact that
$\mathbb{E}\exp(\sigma W_t) = \exp(0.5\sigma^2 t)$:

$$
\mathbb{E}S_{t_i} = \gamma_1 \ldots \gamma_i\, S_{t_0}.
$$

Plugging this into (2.4) one gets

$$\mathbf{EG} = -nk + \sum_{i=1}^{n} \Delta\phi_{t_i} \, \delta_i \, \gamma_1 \ldots \gamma_i \, S_{t_0}.$$

Using (2.3) the variance of G can be written and decomposed as follows:

$$
\begin{aligned}
\operatorname{var}(G) &= \operatorname{var}\left(\sum_{i=1}^{n} \Delta\phi_{t_i} \, \delta_i \, S_{t_i}\right) \\
&= \sum_{i=1}^{n} \left(\Delta\phi_{t_i} \, \delta_i\right)^2 \operatorname{var}\left(S_{t_i}\right) \\
&\quad + 2 \sum_{i<j}^{n} \Delta\phi_{t_i} \, \delta_i \, \Delta\phi_{t_j} \, \delta_j \operatorname{cov}\left(S_{t_i}, S_{t_j}\right).
\end{aligned}
$$

Thus, all that remains is the computation of variances and covariances. Based on the relationships

$$
\begin{aligned}
\operatorname{var}(S_{t_i}) &= \mathbf{E}S_{t_i}^2 - (\mathbf{E}S_{t_i})^2 \\
\operatorname{cov}(S_{t_i}, S_{t_j}) &= \mathbf{E}(S_{t_i} S_{t_j}) - (\mathbf{E}S_{t_i})(\mathbf{E}S_{t_j})
\end{aligned}
$$

it suffices to compute $\mathbf{E}S_{t_i}^2$ and $\mathbf{E}(S_{t_i} S_{t_j})$. Based on (2.12), the second moment can be written as

$$
\begin{aligned}
S_{t_i}^2 &= \gamma_1^2 \ldots \gamma_i^2 \, S_{t_0}^2 \exp\left[-\sigma^2(t_i - t_0)\right] \exp\left[2\sigma(W_{t_i} - W_{t_{i-1}})\right] \\
&\quad \ldots \exp\left[2\sigma(W_{t_1} - W_{t_0})\right].
\end{aligned}
$$

By using $\mathbf{E}\exp(2\sigma W_t) = \exp(2\sigma^2 t)$ one gets

$$\mathbf{E}S_{t_i}^2 = \gamma_1^2 \ldots \gamma_i^2 \, S_{t_0}^2 \exp\left[\sigma^2(t_i - t_0)\right],$$

implying

$$\operatorname{var}(S_{t_i}) = \gamma_1^2 \ldots \gamma_i^2 \, S_{t_0}^2 \left(\exp\left[\sigma^2(t_i - t_0)\right] - 1\right).$$

The covariance terms can be calculated in an analogous fashion. Since

$$
\begin{aligned}
S_{t_i} S_{t_j} &= S_{t_0}^2 \gamma_1^2 \ldots \gamma_i^2 \gamma_{i+1} \ldots \gamma_j \\
&\quad \exp[-1/2\sigma^2(t_i - t_0)] \, \exp[-1/2\sigma^2(t_j - t_0)] \\
&\quad \exp[2\sigma(W_{t_i} - W_{t_{i-1}})] \ldots \exp[2\sigma(W_{t_1} - W_{t_0})] \\
&\quad \exp[\sigma(W_{t_j} - W_{t_{j-1}})] \ldots \exp[\sigma(W_{t_{i+1}} - W_{t_i})]
\end{aligned}
$$

one obtains

$$
\begin{aligned}
\mathbf{E}(S_{t_i}S_{t_j}) &= S_{t_0}^2\gamma_1^2\ldots\gamma_i^2\gamma_{i+1}\ldots\gamma_j \\
&\quad \exp[-1/2\sigma^2(t_i-t_0)]\ \exp[-1/2\sigma^2(t_j-t_0)] \\
&\quad \exp[2\sigma^2(t_i-t_0)]\ \exp[1/2\sigma^2(t_j-t_i)] \\
&= S_{t_0}^2\gamma_1^2\ldots\gamma_i^2\gamma_{i+1}\ldots\gamma_j\ \exp[\sigma^2(t_i-t_0)].
\end{aligned}
$$

So, the covariance can be written as

$$
\mathrm{cov}(S_{t_i}S_{t_j}) = \gamma_1^2\ldots\gamma_i^2\,\gamma_{i+1}\ldots\gamma_j\,S_{t_0}^2\left(\exp\left[\sigma^2(t_i-t_0)\right]-1\right).
$$

References

Almgren, R. and N. Chriss (2000): Optimal execution of portfolio transactions, *Journal of Risk* 3 (Winter 2000/2001), 5–39.

Amihud, Y. and H. Mendelson (1982): Asset price behavior in a dealership market, *Financial Analysts Journal* 42, 50–9.

Bank, P. and D. Baum (2002): Hedging and portfolio optimization in financial markets with a large trader, Humboldt University, Berlin, Germany.

Bertsimas, D. and A. W. Lo (1998): Optimal control of execution, *Journal of Financial Markets* 1, 1–50.

Bessembinder, H. (1994): Bid–ask spreads in the interbank foreign exchange markets, *Journal of Financial Economics* 35, 317–48.

Cuoco, D. and J. Cvitanić (1998): Optimal consumption choices for a large investor, *Journal of Economic Dynamics and Control* 22, 401–36.

Cvitanić, J. and J. Ma (1996): Hedging options for a large investor and forward-backward SDE's, *Annals of Applied Probability* 6, 370–98.

Dubil, R. (2002): Optimal liquidation of large security holdings in thin markets, University of Connecticut, Storrs, USA.

Esser, A. and B. Mönch (2002): Modeling feedback effects with stochastic liquidity, Goethe University, Frankfurt am Main, Germany.

Foster, F.D. and S. Viswanathan (1993): The effect of public information and competition on trading volume and price volatility, *Review of Fiancial Studies* 6, 23–56.

Frey, R. (1998): Perfect option replication for large traders, *Finance and Stochastics* 2, 115–42.

Frey, R. (2000): Market illiquidity as a source of model risk in dynamic hedging, in *Model Risk*, ed. by R. Gibson, RISK Publications, London, 125–36.

Frey, R. and P. Patie (2001): Risk management for derivatives in illiquid markets: A simulation study, RiskLab, ETH-Zentrum, Zürich, Switzerland.

Harris, L.E. (1986): A transaction data study of weekly and intradaily patterns in stock returns, *Journal of Financial Economics* 16, 99–118.

Hasbrouck, J. (1991): Measuring the information content of stock trades, *Journal of Finance* 46, 179–207.

Hisata, Y. and Y. Yamai (2000): Research toward the practical application of liquidity risk evaluation methods, *Monetary and Economic Studies*, December, 83–128.

Holthausen, R. W., R. W. Leftwich, and D. Mayers (1987): The effect of large block transactions on security prices, *Journal of Financial Economics* 19, 237–67.

Huang, R. D. and R. W. Masulis (1999): FX spreads and dealer competition across the 24-hour trading day, *Review of Financial Studies* 12, 61–94.

Jain, P. and G. H. Joh (1988): The dependence between hourly prices and trading volume, *Journal of Financial and Quantitative Analysis* 23, 269–84.

Kampovsky, A. and S. Trautmann (2000): A large trader's impact on price processes, Johannes-Gutenberg-Universität, Mainz, Germany.

Kirchner, T. and C. Schlag (1998): An explorative investigation of intraday trading on the German Stock market, *Finanzmarkt und Portfolio Management* 12, 13–31.

Lin, J., G. C. Sanger, and G. G. Booth (1995): Trade size and components of the bid–ask spread, *Review of Fiancial Studies* 8, 1153–84.

Linnainmaa, J. (2003): Who makes the limit order book? Implications for contrarian strategies, attention-grabbing hypothesis, and the disposition effect, University of California, Los Angeles, USA.

Liu, H. and J. Yong (2004): Option pricing with an illiquid underlying asset market, Washington University, USA and University of Central Florida, Orlando, USA.

Lyons, R. K. (1995): Tests of microstructural hypotheses in the foreign exchange market, *Journal of Financial Economics* 39, 321–51.

McInish, T.H. and R. A. Wood (1990): A transaction data analysis of the variability of common stock returns during 1980–1984, *Journal of Banking and Finance* 14, 99-112.

McInish, T. H. and R. A. Wood (1992): An analysis of intraday patterns in bid–ask spreads for NYSE stocks, *Journal of Finance* 47, 753–64.

Ranaldo, A. (2000): Intraday trading activity on financial markets: The Swiss evidence, Ph.D. thesis, Université de Fribourg, Switzerland.

Schönbucher, P. J. and P. Wilmott (2000): The feedback effect of hedging in illiquid markets, *SIAM Journal of Applied Mathematics* 61(1), 232–72.

Sircar, K. R. and G. Papanicolaou (1998): General Black–Scholes models accounting for increased market volatility from hedging strategies, *Applied Mathematical Finance* 5, 45–82.

Wei, P. H. (1992): Intraday variations in trading activity, price variability and the bid–ask spread, *Journal of Financial Research* 15, 265–76.

Wood, R. A., T. H. McInish, and J. K. Ord (1985): An investigation of transaction data for NYSE stocks, *Journal of Finance* 40, 723–39.

3 The Navigation of an Iceberg: The Optimal Use of Hidden Orders

3.1 Introduction

The rapid development in technology over the last couple of years has permitted many stock exchanges to transfer trading from open outcry markets, where market makers or specialists act as intermediaries, to screen-based electronic markets. Typically, electronic trading platforms provide market participants with information on an anonymous open order book during continuous trading in real time. Usually the limits, the accumulated order volumes of each limit, and the number of orders in the book at each limit are displayed, so that traders can assess the altering order flow and the market liquidity.

What does the existence of an open order book imply for investment firms who want to submit limit orders, the total volume of which is large relative to others in the market? No doubt, exposing large limit orders in an open order book may reveal the investor's motives for trading and may raise suspicion that the originator of the large order has access to private information about the true value of the security under consideration. Consequently, other market participants change their own order submission strategy, which in turn lowers the probability that the large order will be executed at the prespecified limit. The investment firm then has to choose a less favorable limit if it wants to increase the probability of execution and thus suffers losses from the *indirect* adverse price impact of its large exposure in the order book. A possible solution is not to submit one large limit order but to split the order into several smaller limit orders, which are submitted over time. For this reason many electronic trading platforms introduced so-called iceberg orders. Euronext, the Toronto Stock Exchange, the London Stock Exchange (with its order-driven services SETS, SETSmm, and IOB), and XETRA are just some prominent examples. Iceberg orders allow market participants to submit orders with only a certain portion of the order publicly disclosed. The metaphor alludes to the fact that in nature an iceberg's biggest part floats unobservable under the water. Only one-ninth of the mass of ice is seen above the water surface.

An iceberg order is specified by its mandatory limit, its overall volume, and a peak volume. The peak is the visible part of the iceberg order and is introduced into the order book with the original time stamp of the iceberg order according to price/time priority. As soon as the disclosed volume of an iceberg order has received a complete fill and a hidden volume is still available, a new peak is entered into the book with a new time stamp. The new peak behaves in an identical manner to a conventional limit order. From this point of view a pure limit order is basically a special case of an iceberg order where the peak volume coincides with the total order volume.

However, it is important to note that iceberg orders exhibit a less favorable time priority compared with pure limit orders. After the peak of an iceberg order is completely matched, all visible limit orders at the same limit that were entered before the new peak is displayed take priority, i.e. they must have received a complete fill before the new peak comes to execution.

When submitting an iceberg order to the market, several issues have to be considered. Imagine, for example, that the management of a mutual fund has to close a large position in a single stock within one trading day. Using an iceberg order with only a small peak size allows it to minimize the adverse informational impact of disclosing the actual order volume. However, the smaller the peak size the less favorable the time priority of the overall order. Thus, choosing a peak size that is too small seems suboptimal. Such a strategy would significantly lengthen the time to complete execution or would make a complete fill unlikely. Moreover, the fund managers have to choose a reasonable limit for the order. If the limit is too low, one may miss some trade opportunities, i.e. one would give away the chance to participate in raising stock prices. Otherwise, if the limit is too ambitious, the order is unlikely to receive a complete fill.

In this chapter this tradeoff is modeled analytically in a continuous time setup where a large position in a single stock is to be liquidated within a finite trading window.[1] We assume that the investor uses an iceberg order and follows a static strategy, i.e. once the limit and the peak size of the iceberg order are chosen, the trader sticks to this strategy over a fixed period. We then determine the optimal peak size and the optimal order limit by maximizing the expected payoff of the liquidation strategy under certain assumptions concerning the execution risk of the iceberg order. Note that a pure limit order would be also an admissible solution to our optimization problem.

Unless an iceberg sell order is immediately executable, i.e. the limit is so low that it is actually a market sell order, the probability of receiving a complete fill within a finite time horizon is strictly smaller than one. In principle at least two alternative approaches would be able to incorporate execution risk into a liquidation model.

First, one may assume that the investor is forced to trade the remaining shares with a market order if the iceberg order fails to receive complete execution. We call this setup the *self-contained approach*. Market orders are executed immediately. They use liquidity from the book until they are completely filled. Consequently the investor has to bear a liquidity discount, so that he or she gets penalized for every share that could not be sold via the iceberg order. However, in our opinion such a rigorous assumption may

---

[1]The analysis for a purchasing strategy is symmetric.

not always be justified in practice, especially if the remaining order volume under consideration cannot be absorbed by the market without a significant price change.

In this case, investors typically follow an adaptive strategy, i.e. they review their orders frequently and adjust them if the market moves away from the prespecified order limit. For this reason we also propose a different approach that considers the execution probability as a boundary condition, i.e. only those combinations of peak size and limit are admissible that ensure a certain execution probability within a prespecified time horizon. We call this model the *open approach.*

Compared with the first one the latter framework is rather flexible and does not require any assumption concerning the liquidation of the unexecuted part of the iceberg order. To get a flavor of the concept, imagine, for example an investor who wants to liquidate a large position, say, within one week. At the end of each trading day the investor inspects the state of the iceberg order and, if necessary, adjusts the limit or the peak size to reach the target.[2]

The *open approach* can assist the investor in this procedure. It deals with the optimal combination of order limit and peak size that maximizes the expected liquidation revenues in the case of complete execution, given that the probability to receive a complete fill exceeds a certain level, for example 40% within one trading day. If the order remains partially or completely unexecuted by the end of the first day, the investor may wish to rerun the optimization at the second day and thereby increase the execution probability, let's say, to 60% and so on. If a substantial part of the order is still unexecuted on the last day of the week, the investor will probably choose a minimum execution probability that is close to one. In principle one can also specify a utility function for the investor to model the trade-off between

---

[2]Note that at some exchanges unexecuted iceberg orders are deleted automatically by the system at the end of each trading day and must be resubmitted if desired by the investor on the next trading day. In this case a daily adjustment of order limit and peak size seems very plausible.

expected payoffs and execution risk. However, in order to keep the problem tractable for exposition we will not address this issue in this chapter.

We present a theoretical framework for both the *open* and the *self-contained approach*. Although the underlying assumptions of the latter model are certainly questionable from an empirical point of view we believe that its basic structure may serve as a guideline to build more sophisticated models, for example by implementing an individual penalty function for the unexecuted part of the iceberg order that meets the specific requirements of the investor under consideration. The numerical analysis that illustrates the theoretical part will focus on the *open approach*.

The technical design of the model can be summarized as follows: During continuous trading a transaction takes place if an order becomes executable against orders on the other side of the book. Thus, for an iceberg sell order that is stored on the ask side of the book the dynamics of the best bid price are of special interest. We model the best bid price as a stochastic process in continuous time and assume a constant best bid size. If the stochastic process hits the limit of the iceberg order a transaction is executed and the stochastic process jumps back to the next lower limit. Whether the peak of the iceberg order or another sell order at the same limit is processed at this event depends on the relative time priority of the orders. If new orders with the same limit as the iceberg order are submitted continuously to the book the time priority of the iceberg order deteriorates compared with a pure limit order. The smaller the peak size of the iceberg order, the more often the limit must be hit such that the iceberg order receives a complete fill.

On the other hand, a smaller peak size lowers the adverse informational impact of showing the actual order volume in an open book. We model the drift of the stock price process as a function of the visible order imbalance. When the peak size of an iceberg order enters the book the visible order imbalance changes. We define the order imbalance as the total visible order volume (in number of shares) stored on the bid side of the order book divided by the total visible order volume stored on the bid side and on the ask side of the order book. We exemplify empirically, using order book data, that current variations in the visible order book imbalance are positively correlated with future returns. Thus, the higher the peak size of an iceberg sell order, the smaller the order imbalance and the smaller the expected returns in the next time intervals. Consequently, a higher peak size results in a smaller probability that the stock price process will reach the prespecified limit within the given time horizon.

In total, one can observe two opposite effects if the peak size of an iceberg sell order is reduced in our model:

- The drift of the stochastic process is reduced to a smaller extent when the order enters the book.

- The number of times the limit must be hit in order to process the iceberg order completely increases.

While the first effect is beneficial for the originator of the iceberg order, the latter is not. The proposed framework weights these effects and identifies the optimal combination of peak size and order limit.

The rest of the chapter is organized as follows: Section 3.2 briefly reviews the related literature. The dataset used to exemplify the theoretical ideas throughout this chapter is described in Section 3.3. Section 3.4 introduces the theoretical setup for both the *self-contained* and the *open approach*. In Section 3.5 we explicitly model the drift as a function of the order imbalance. The open approach to determine the optimal combination of order limit and peak size is calibrated with a clinical order book data sample in Section 3.6 so that one can get an impression of the optimal strategies for different scenarios. The chapter concludes in Section 3.7 with a brief summary and a discussion of issues for further research.

3.2 Related Literature

A number of empirical studies shed light upon the use of hidden orders[3] and the associated motives of traders.

Aitken, Brown, and Walter (1996) state that approximately 6% of orders accounting for 28% of shares traded at the Australian Stock Automated Trading System (SEATS) were undisclosed in 1993. Aitken, Berkman, and Mak (2001) find that undisclosed limit orders are used to reduce the option value of limit orders. This follows the intuition that limit orders can become mispriced when new public information arrives. Some authors, for example Copeland and Galai (1983), therefore characterize limit buy (sell) orders as free put (call) options to other market participants. Pardo and Pascual (2003) use six months of limit order book and transaction data on 36 stocks from the Spanish Stock Exchange (SSE) and report that 26% of all trades (20% of all non-aggressive trades and 42% of all aggressive trades)[4] involve hidden volume. They provide evidence that liquidity suppliers use iceberg orders to mitigate adverse selection costs if new information is released to the market, and that hidden orders temporarily increase the aggressiveness of traders when revealed to the public. D'Hondt, De Winne, and François-Heude (2003) investigate data for six CAC40 stocks traded at Euronext and show that 30% of the depth is hidden in the whole book. The authors highlight that hidden orders are more frequently canceled than usual orders, that iceberg orders are less likely to be totally filled and that the limit of hidden orders is modified more often than that of pure limit orders.

The modeling of optimal liquidation strategies attracts more and more attention by researchers. Bertsimas and Lo (1998), Almgren and Chriss (2000), Hisata and Yamai (2000), Dubil (2002), and Mönch (2003) investigate liqui-

---

[3]Note that at some exchanges the expression "hidden order" is reserved exclusively for orders that are *completely* invisible to other market participants. However, as it is common practice in the literature that is related to this chapter we use it as a synonym for an iceberg order.

[4]A buyer- or seller-initiated trade is defined as *aggressive* if it consumes, at least, the best quote on the opposite side of the book.

dation strategies for large security positions if market orders are employed as trading instruments. The papers differ from each other mainly in the definition of the stock price dynamics, the modeling of the price impact function, and whether the final time horizon is given exogenously or modeled endogenously.

Compared with market orders, the analysis of optimal liquidation strategies for limit orders is more complex. While the former are matched immediately (provided a sufficient market depth) the execution probability of the latter order type depends critically on the respective limit. Wald and Horrigan (2001) estimate a probit model that characterizes the execution probability depending on a number of variables as the order limit, subsequent realized returns, the bid–ask spread and so on. Lo, MacKinlay, and Zhang (2002) compare empirically three different approaches to determine the execution probability of limit orders using order book data for the 100 largest stocks in the S&P 500 from August 1994 to August 1995. First, they model the execution of a limit order as the first passage time of a geometric Brownian motion to the limit price and find that the predictive power of this setup is only moderate. The first passage time model suffers from important shortcomings. It neither considers the time priority, the order size, a potential adverse impact of revealing large limit orders in the book nor does it distinguish between time-to-first-fill and time-to-completion. As mentioned above, these limitations are eased in the framework proposed in this chapter. Second, the authors consider first-passage times determined by historical time series of transaction data. As this approach also ignores the time priority and current market conditions it is not able to represent actual limit order execution times adequately. Finally, the article proposes an econometric model of limit order execution times based on survival analysis and actual limit order data. This empirical approach is a *reduced form* model as it leaves open which mechanism actually causes the execution of a limit order. The model uses eight explanatory variables that are updated in real time to capture current market conditions and three explanatory variables that are updated monthly to model differences across stocks. The authors make some assumptions that may not always be justified empirically to keep the framework tractable. For example the authors assume time-independent covariates. Nevertheless, the empirical framework is able to predict actual limit order executions very well. However, the model requires a continuous update with order book data. Thus, it may be only of limited use if such data are not available in real time.

The intention of our framework is closely related to that of Lo, MacKinlay, and Zhang (2002) although we use a *structural* approach to model explicitly the functionality of the order book. Both approaches focus on the modeling of the execution probability of limit or, in our case, more generally, iceberg orders. In our approach this probability is obtained endogenously and in the parametric model of Lo, MacKinlay, and Zhang (2002) the form of the probability density function is specified exogenously. While Lo, MacKinlay, and Zhang (2002) just present a framework to estimate the time-to-first-fill and the time-to-complete execution of a limit order, our approach is more flexible as it is able to capture every state of partial execution. Furthermore, we extend the analysis to the identification of optimal liquidation strategies for different scenarios.

Cho and Nelling (2000) estimate the execution probability of a limit order conditional on order-specific variables and other variables that capture general market conditions using quote data for 144 NYSE stocks from November 1990 to January 1991. The authors observe that the longer a limit order is outstanding, the less likely it is to receive a complete fill. Furthermore, they find that the execution probability is low when the limit price is far away from the current quote, when the order volume is high, when spreads are narrow, and when volatility is low.

The hypothesis that the order imbalance is a proxy for the execution probability of limit orders and influences the order submission strategy of investors is supported by many authors. Chordia and Subrahmanyam (2002) analyze time series of daily order imbalances and individual stock returns for the period 1988–98 using a comprehensive sample of NYSE stocks. They find that lagged imbalances bear a positive predictive relation to current day returns. Furthermore, they observe that daily imbalances are positively autocorrelated. Ranaldo (2004) uses data of 15 stocks quoted on the Swiss Stock Exchange. He finds that orders are submitted more (less) aggressively when the outstanding order volume on the same (opposite) side of the book is large. Furthermore, he observes that buyers are more concerned about the opposite side of the book, while sellers are more concerned about their own side. Parlour (1998) presents a dynamic setup of an order book, where investors anticipate that the own order placement strategy influences the following order flow and where the execution probability of limit orders is modeled endogenously.

3.3 Description of the Market and Dataset

The functionality of the liquidation model proposed in this chapter will be illustrated by using the order book data sample for the company MAN that was introduced in Section 2.3.

In this dataset we counted 786 iceberg sell orders, 140,948 pure limit sell orders, and 4,130 market sell orders. At first sight the hidden part of the order book seems tiny. However, analyzing the average volume of each trading instrument more deeply changes this impression slightly. Iceberg orders exhibit an average volume of 16,037 shares, whereas pure limit and market orders just have an average order volume of 964 and 1,069 shares. Due to this fact, hidden orders represent a remarkable proportion of 8.24% of the overall volume on the ask side of our sample order book. Pure limit sell orders and market sell orders provide 88.87% and 2.89% of the liquidity on the ask side.

Figure 3.1 shows the spectrum of the observed initial volumes of all hidden sell orders. In Figures 3.2 and 3.3 we address the issue of how market participants choose the limit and the peak size in practice. Obviously there exists a strong preference to specify a peak size that corresponds to a tenth of the overall order volume, as one can see in Figure 3.2. Roughly 37% of all market participants follow such a strategy. To investigate the difference between the chosen limit and the current best bid price at the time of submission

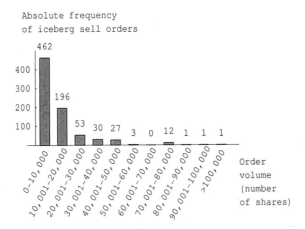

Fig. 3.1: Absolute frequencies of order volumes of all observed iceberg sell orders in the sample.

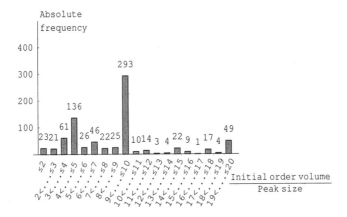

Fig. 3.2: Ratio of initial order volume to peak size of all different iceberg sell orders in the sample.

we consider only the 702 iceberg sell orders that entered the book during
continuous trading. Figure 3.3 shows that the majority of market participants
set the limit between 5 and 15 cents above the best bid price. Note that the
average bid–ask spread in our sample is 7 cents and the average midprice
€ 26.91.

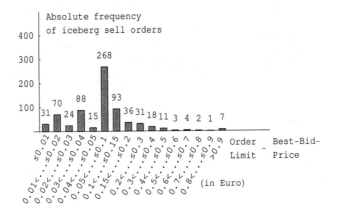

Fig. 3.3: Absolute price difference between the chosen limit and the current best
bid price at time of entry of all observed iceberg sell orders in the sample
entered during continuous trading.

With respect to the success of the observed trading strategies, Table 3.1
delivers an insight into the empirical execution probability of hidden orders.
Less than 18% of all iceberg sell orders were executed completely. Almost
30% of all iceberg orders received a partial fill before expiry or cancelation by
the investor. The majority (52%) of all hidden sell orders were canceled or
expired completely unexecuted. Looking at the median of the observed time
between entry and complete execution or deletion one can state that market
participants check the state of their orders frequently and cancel them if
prices move away from the limit.

Table 3.1: Execution or deletion of iceberg sell orders in the order book sample.

	Number of iceberg sell orders	Average ratio of executed to initial volume; median in parentheses	Average time between entry and complete execution or deletion (hh:mm:ss); median in parentheses
Completely executed	139	100.00% (100.00%)	00:40:48 (00:10:46)
Partially executed	231	32.44% (28.00%)	00:50:04 (00:09:29)
Completely unexecuted	416	0.00% (0.00%)	03:32:43 (00:09:45)

3.4 The Basic Setup

3.4.1 General Idea and Dynamics

This section introduces the basic concepts and provides the motivation for the assumptions that have been made. Assume that the large investor holds ϕ_0 shares that should be liquidated before time T. For this purpose the trader submits an iceberg sell order that is stored on the ask side of the order book. The investor assigns a peak size ϕ_p and a limit $\bar{S}$ to the iceberg order. The latter is strictly higher than the initial best bid price S_0 such that the first proportion of the order is not immediately executable.

The best bid price S_t is modeled by a kind of jump-diffusion process. For $S_t < \bar{S}$ it follows a geometric Brownian motion:

$$dS_t = \mu S_t \, dt + \sigma S_t \, dW_t \qquad \text{with } S_0 < \bar{S}. \tag{3.1}$$

Throughout this section the drift μ is assumed to be a constant. In Section 3.5 we ease this restriction and model the drift as a function of the chosen peak size.

When the process hits the limit of the iceberg order, i.e. $S_{t-} = \bar{S}$, a small downward jump to the next order book entry on the bid side occurs such that $S_t = (1 - \varepsilon)\,\bar{S}$. For sake of simplicity the jump size is modeled as a constant throughout this chapter. Figure 3.4 illustrates the general setting described so far.

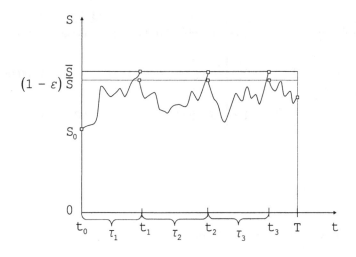

Fig. 3.4: If the best bid price S_t hits the limit of the iceberg order $\bar{S}$, a small downward jump to the next limit on the bid side, i.e. to $(1 - \varepsilon)\,\bar{S}$, occurs.

Each time the limit $\bar{S}$ is hit by the best bid price a transaction is executed. The transaction size ϕ_s is assumed to be constant over time. Furthermore, we assume that whenever a new tranche of the iceberg order enters the book a fixed volume ϕ_a of other sell orders that exhibit a better time priority is already stored at the same limit. These orders must be matched before the current peak of the iceberg order becomes executable. Thus, one can observe the following sequence of newly displayed order quantities over time: $\phi_a, \phi_p, \phi_a, \phi_p$, and so on. Table 3.2 summarizes the notation that is used throughout the rest of this chapter.

Table 3.2: Notation used in the proposed liquidation model.

Variable	Meaning
ϕ_0	Total number of shares that have to be liquidated by the large investor before time T
$\phi_p (\leq \phi_0)$	Assigned peak size to the iceberg order
ϕ_a	Volume of other sell orders at limit $\bar{S}$ that is already stored on the ask side when a fresh peak enters the order book
ϕ_s	Total transaction volume that is processed each time the limit is hit

The objective of the investor is the identification of the optimal combination of peak size ϕ_p and limit $\bar{S}$. The time horizon is T, and the investor is interested in the expected payoff of his or her liquidation strategy. To deal with execution risk, we consider two alternative approaches. In the *open approach*, the investor chooses a lower bound P^* for the probability that the submitted iceberg order receives a complete fill up to time T. For the *self-contained approach* we assume that if the iceberg order is not completely executed before time T, the large investor submits a market order to sell the remaining part of the shares. Consequently, the trader has to bear a significant liquidity discount, denoted by $\Psi \cdot S_T$, where $\Psi \gg \varepsilon$. Thus, the investor will receive $(1 - \Psi) S_T$ for each of the remaining shares. It seems reasonable to model Ψ as a function of the number of shares that are sold by submitting a market order at T.

The remaining part of this section is dedicated to the derivation of the formulas necessary to implement the *open* and the *self-contained approach*. Obviously, the liquidation value depends on the actual number of times the best bid price hits the prespecified limit of the iceberg order, which in turn is a random variable in the proposed setup. Thus, in Subsection 3.4.2 we introduce formulas to compute the executed volume of the iceberg order conditional on the event that the limit is hit a certain number of times. In Subsection 3.4.3 we calculate the probability that the limit is hit a certain number of times. Furthermore, in this subsection we present the objective function of the investor for both the *open* and the *self-contained approach*.

3.4.2   Execution of the Iceberg Order

The number of times n^* the limit $\bar{S}$ must be hit by the best bid price such that the iceberg order is completely satisfied is given by

$$n^* = \left\lceil \frac{(\phi_0/\phi_p)\phi_a + \phi_0}{\phi_s} \right\rceil = \left\lceil \frac{\phi_0(1 + \phi_a/\phi_p)}{\phi_s} \right\rceil.$$

The brackets $\lceil . \rceil$ are called upper Gaussian brackets, with $\lceil x \rceil = \min\{z \in \mathbb{Z} : z \geq x\}$ and $\mathbb{Z}$ as the set of integers. Note that $n^* \cdot \phi_s$ corresponds to the total order volume that is matched after n^* transactions, whereas ϕ_0 shares originate from the iceberg order and $(\phi_0/\phi_p)\phi_a$ shares from other orders.

After the limit is hit n times the number of executed shares of the iceberg order is given by

$$h(n) := \min \left\{ \max \left[\phi_p \left\lfloor \frac{n\phi_S}{\phi_a + \phi_p} \right\rfloor, \right. \right.$$

$$\left. \left. n\phi_S - \phi_a \left(1 + \left\lfloor \frac{n\phi_S}{\phi_a + \phi_p} \right\rfloor \right) \right], \phi_0 \right\},$$

where $\lfloor x \rfloor = \max\{z \in \mathbb{Z} : z \leq x\}$. As long as $n \leq n^*$ the first element of the max-expression is larger than the second if other sell orders at the same limit exhibit a better time priority than the current peak of the iceberg order. If the current peak of the iceberg order takes time priority over other sell orders at the same limit, the second term in the max-expression is larger than the first.

The outstanding order volume of the iceberg order after the n-th hit of the limit is then equal to

$$\phi_0 - h(n).$$

3.4.3 Liquidation Value

Armed with the results of the previous section one can now calculate the liquidation value conditional on the event that the limit is hit a certain number of times. Let M denote the number of times the limit is hit *before* time T. If $M \geq n^*$, the liquidation value G is given by $G = \phi_0 \bar{S}$, since the iceberg order is completely executed at time T. The *open approach* simply maximizes this expression by solving the following optimization problem:

$$\max_{\{\phi_p, \tilde{S}\}} \quad \phi_0 \bar{S}$$

$$s.t. \quad \begin{aligned} P^* &\leq P(M \geq n^*) \\ S_0 &< \bar{S} \\ \phi_p &\leq \phi_0. \end{aligned}$$

In contrast to the *open approach*, which focuses on the case of full execution, the *self-contained approach* considers also those states of the world where $M < n^*$. In this setup, if $M = n$, the trader receives $h(n)\bar{S}$ for the executed part of the iceberg order and $[\phi_0 - h(n)][1 - \Psi(\phi_0 - h(n))]S_T$ for the remaining part that is liquidated using a market order at time T. Given the realizations of M and S_T the liquidation value can be calculated by

$$G_{M=n} = h(n)\bar{S} + [\phi_0 - h(n)][1 - \Psi(\phi_0 - h(n))]S_T.$$

However, at time t_0, both M and S_T are random variables. Thus, in order to derive the expected liquidation value one has to weight all possible realizations of the liquidation value G_M by their probabilities, whereas S_T depends on the realization of M. Thus, the expected liquidation value can be written as

$$\begin{aligned} \mathbf{E}G = {} & \sum_{n=0}^{\infty} P(M = n) \times \\ & \left\{ h(n)\bar{S} + [\phi_0 - h(n)][1 - \Psi(\phi_0 - h(n))] \times \right. \\ & \left. \mathbf{E}(S_T | M = n) \right\} \end{aligned}$$

$$= \mathrm{P}(M = 0)\, \phi_0 \cdot [1 - \Psi(\phi_0)]\, \mathbf{E}\,(S_T|\, M = 0)$$

$$+ \sum_{n=1}^{n^*-1} \mathrm{P}(M = n) \left\{ h\,(n)\,\bar{S} + [\phi_0 - h\,(n)] \times \right.$$

$$\left. [1 - \Psi\,(\phi_0 - h\,(n))]\, \mathbf{E}\,(S_T|\, M = n) \right\}$$

$$+ \mathrm{P}(M \geq n^*)\, (\phi_0\, \bar{S})\,.$$

For the *self-contained approach* we need to solve the following optimization problem:

$$\max_{\{\phi_p, \tilde{s}\}} \quad \mathbf{E}G$$

$$s.t. \quad S_0 \; < \; \bar{S}$$

$$\phi_p \; \leq \; \phi_0.$$

It remains to calculate the following quantities:

- $\mathrm{P}(M = 0)$

- $\mathrm{P}(M = n)$, for $n = 1, \dots, n^* - 1$

- $\mathrm{P}(M \geq n^*)$

- $\mathbf{E}\,(S_T|\, M = 0)$

- $\mathbf{E}\,(S_T|\, M = n)$, for $n = 1, \dots, n^* - 1$.

For this purpose we must compute the distributions of the hitting times, denoted by t_i, $i = 1, \dots, n$. The time periods between two successive hitting times will be denoted by $\tau_i := t_i - t_{i-1}$, $i = 2, \dots, n$ and we will let $\tau_1 = t_1$.

The distribution of τ_1 can be calculated as follows: Since the process for the best bid price S_t up to τ_1 follows a geometric Brownian motion, see equation (3.1), the logarithm of the process is an arithmetic Brownian motion

$$\mathrm{d}\,(\ln S_t) = \left(\mu - \sigma^2/2\right) \mathrm{d}t + \sigma \mathrm{d}W_t.$$

Note that if an arithmetic Brownian motion has a negative drift, i.e. when $\mu < \sigma^2/2$ in our model, then τ_1 has a *defective* density function whose integral over $[0, \infty)$ is less than one, see, e.g., Karlin and Taylor (1975), p. 362. Thus, the probability that the limit will *never*(!) be reached is positive. Specifically, the probability for the event $\{\tau_1 < \infty\}$ is given by

$$P\left(\tau_1 < \infty\right) = \begin{cases} 1 & \text{for } \mu \geq \sigma^2/2 \\ \exp\left[-2\ln\left(\bar{S}/S_0\right)|\mu - \sigma^2/2|/\sigma^2\right] & \text{for } \mu < \sigma^2/2. \end{cases} \quad (3.2)$$

The distribution of the first hitting time of $\ln \bar{S}$, starting at $\ln S_0 < \ln \bar{S}$, conditional on the event $\{\tau_1 < \infty\}$, is given by

$$\tilde{f}_{0,1}(t) = \frac{\ln\left(\bar{S}/S_0\right)}{\sigma\sqrt{2\pi t^3}} \exp\left\{-\frac{\left[\ln\left(\bar{S}/S_0\right) - |\mu - \sigma^2/2|\,t\right]^2}{2\sigma^2 t}\right\}. \quad (3.3)$$

Taking the product of (3.2) and (3.3) we can write the unconditional (defective) density of τ_1 as

$$\begin{aligned} f_{0,1}(t) &= \tilde{f}_{0,1}(t)\,P\left(\tau_1 < \infty\right) \\ &= \frac{\ln\left(\bar{S}/S_0\right)}{\sigma\sqrt{2\pi t^3}} \exp\left\{-\frac{\left[\ln\left(\bar{S}/S_0\right) - (\mu - \sigma^2/2)\,t\right]^2}{2\sigma^2 t}\right\}. \quad (3.4) \end{aligned}$$

At τ_1 the process independently restarts at $\bar{S}(1 - \varepsilon)$, following again a geometric Brownian motion. To derive the (defective) density of the first hitting time *after the restart* (denoted by $\tau_2 = t_2 - t_1$) we just need to replace $\ln S_0$ by $\ln\left[\bar{S}(1 - \varepsilon)\right]$ in equation (3.4) if we assume a constant drift μ. Thus, for $n^* \geq 2$ one can write

$$f_{n-1,n}(t) = \frac{-\ln\left(1 - \varepsilon\right)}{\sigma\sqrt{2\pi t^3}} \exp\left\{\frac{-\left[-\ln\left(1 - \varepsilon\right) - (\mu - \sigma^2/2)\,t\right]^2}{2\sigma^2 t}\right\}. \quad (3.5)$$

Since t_2 can be decomposed into the sum of the two independent random variables τ_1 and τ_2, i.e. $t_2 = \sum_{i=1}^{2} \tau_i$, the (defective) density $f_{0,2}$ of t_2 is simply the convolution of the corresponding (defective) densities, given by

$$f_{0,2}(t) \equiv (f_{0,1} \star f_{1,2})(t) := \int_0^t f_{0,1}(t - u)f_{1,2}(u)du.$$

One can proceed by iterating this methodology: At τ_i, $i \geq 2$ the process independently restarts at $\bar{S}(1 - \varepsilon)$ following a geometric Brownian motion. Thus, the distribution of $t_n = \sum_{i=1}^{n} \tau_i$ is given by

$$f_{0,n}(t) \equiv (f_{0,n-1} \star f_{n-1,n})(t) = (f_{0,1} \star f_{1,2} \star \ldots \star f_{n-1,n})(t). \quad (3.6)$$

Now, we are able to derive the corresponding probabilities by rewriting the number of hits in terms of hitting times. Since the events

$$\{M = n\} \quad \text{and} \quad \{(t_n < T) \wedge (t_{n+1} \geq T)\}$$

are identical, the desired probability for $(M \geq 1)$ can be written as

$$
\begin{aligned}
\mathrm{P}\,(M = n) \;&\equiv\; \mathrm{P}\Big((t_n < T) \wedge (t_{n+1} \geq T)\Big) \\
&=\; \mathrm{P}\,(t_n < T) - \mathrm{P}\,(t_{n+1} < T) \\
&=\; \mathrm{P}\left(\sum_{i=1}^{n} \tau_i < T\right) - \mathrm{P}\left(\sum_{i=1}^{n+1} \tau_i < T\right) \\
&=\; \int_0^T f_{0,n}(t)\,\mathrm{d}t - \int_0^T f_{0,n+1}(t)\,\mathrm{d}t.
\end{aligned}
$$

The probability for the event that the limit is not hit before T is given by

$$
\begin{aligned}
\mathrm{P}\,(M = 0) \;&\equiv\; \mathrm{P}\,(t_1 > T) \\
&=\; 1 - \int_0^T f_{0,1}\,(t)\,\mathrm{d}t.
\end{aligned}
$$

The probability for a complete fill of the iceberg order before T can be computed via

$$
\begin{aligned}
\mathrm{P}\,(M \geq n^*) \;&\equiv\; \mathrm{P}\,(t_{n^*} \leq T) \\
&=\; \int_0^T f_{0,n^*}\,(t)\,\mathrm{d}t.
\end{aligned}
$$

Now the expected liquidation value, conditional on the event that the limit is hit n times before time T, can be calculated. To simplify the explanation, assume for a moment that the hitting times are deterministic. This assumption will be relaxed later. In this case the expression

$$\mathbb{E}\,(S_T |\, M = n)$$

is equal to

$$\mathbf{E}_{t_n}\left(S_T \,\middle|\, \max_{t_n < u < T}(S_u) < \bar{S}\right)$$

$$= \mathbf{E}_{t_n}\left(\exp\left[\ln(S_T/S_{t_n})\right] S_{t_n} \,\middle|\, \max_{t_n < u < T}\left[\ln(S_u/S_{t_n})\right] < \ln(\bar{S}/S_{t_n})\right)$$

$$= \int_{-\infty}^{\ln(\bar{S}/S_{t_n})} \exp(s)\, S_{t_n} g\left(s \,\middle|\, \max_{t_n \le u \le T}\left[\ln(S_u/S_{t_n})\right] < \ln(\bar{S}/S_{t_n}),\, n\right)\mathrm{d}s,$$

since the n-th hit occurs at t_n and the process independently restarts at t_n following a geometric Brownian motion conditional on the event that the threshold $\bar{S}$ is not hit within the time interval from t_n to T.

One may notice that this scenario is similar to the evaluation of knock-out barrier options (see, e.g., Zhang (1998), pp. 203–259). The formula for the conditional density g of $\ln(S_T/S_{t_n})$ is given in Appendix 3.8.

However, for $M = n \ge 1$, t_n is in fact a random variable. Thus, we need to consider the distribution of t_n, conditional on the event $\{t_n \le T \wedge t_{n+1} > T\}$. Due to the independence and identical distributions of τ_i for $i \ge 2$ this conditional density is given by

$$f^{cond}(t)\mathrm{d}t := \mathrm{P}(t_n \in (t, t+\mathrm{d}t)|\tau_{n+1} > T - t)$$

$$= \frac{\mathrm{P}\left((t_n \in (t, t+\mathrm{d}t)) \wedge (\tau_{n+1} > T - t)\right)}{\mathrm{P}(\tau_{n+1} > T - t)}$$

$$= \frac{f_{0,n}(t)\mathrm{P}(\tau_{n+1} > T - t)\mathrm{d}t}{\mathrm{P}\left((t_n \le T) \wedge (t_{n+1} > T)\right)}$$

$$= \frac{f_{0,n}(t)\left(1 - \int_0^{T-t} f_{n,n+1}(s)\mathrm{d}s\right)\mathrm{d}t}{\int_0^T f_{0,n}(u)\,\mathrm{d}u - \int_0^T f_{0,n+1}(u)\,\mathrm{d}u}.$$

Armed with this result we are able to write the conditional expectation of S_T as

$$\mathbf{E}(S_T|M = n)$$

$$= \int_0^T f^{cond}(t)\, \mathbf{E}_t\left(S_T \,\middle|\, \max_{t_n < u < T}(S_u) < \bar{S}\right)\mathrm{d}t$$

$$= \int_0^T \frac{f_{0,n}(t) \left(1 - \int_0^{T-t} f_{n,n+1}(s)ds\right)}{\int_0^T f_{0,n}(u)\,du - \int_0^T f_{0,n+1}(u)\,du} \times$$

$$\left[\int_{-\infty}^{\ln(\bar{S}/S_{tn})} \exp(s)\, S_t \times \right.$$

$$\left. g\left(s \mid \max_{t \le u \le T}[\ln(S_u/S_t)] < \ln(\bar{S}/S_t),\ n\right) ds \right] dt. \qquad (3.7)$$

Note that the integral with respect to t in equation (3.7) has a singularity at the upper end point of the integration range. Thus, for numerical integration one should use a quadrature routine that can handle functions with end-point singularities.[5]

Conditional on the event that the limit is not hit before T, the conditional expectation of S_T simplifies to

$$\mathbf{E}\left(S_T \mid M = 0\right)$$

$$= \int_{-\infty}^{\ln(\bar{S}/S_0)} \exp(s)\, S_0\, g\left(s \mid \max_{0 \le u \le T}[\ln(S_u/S_0)] < \ln(\bar{S}/S_0),\ n = 0\right) ds.$$

The general setup of the alternative approaches is summarized in the following two propositions:

Proposition 3.1 *The **open approach** to determine the optimal combination of the peak size and the limit of an iceberg order can be represented by the following optimization problem:*

$$\max_{\{\phi_p, \bar{s}\}} \quad \phi_0\, \bar{S}$$

$$s.t. \quad P^* \le \int_0^T f_{0,n^*}(t)\, dt$$

$$S_0 < \bar{S}$$

$$\phi_p \le \phi_0,$$

where P^* is given exogenously.

---

[5]For example, imsl_d_int_fcn_sing from the IMSL C-Library is such a routine.

Proposition 3.2 *The **self-contained approach** to determine the optimal combination of the peak size and the limit of an iceberg order can be represented by the following optimization problem:*

$$
\begin{aligned}
\max_{\{\phi_p, \bar{S}\}} \quad & \mathbf{EG} \\
\text{s.t.} \quad S_0 \ &< \ \bar{S} \\
\phi_p \ &\leq \ \phi_0,
\end{aligned}
$$

where $\mathbf{EG}$ *is given by*

$$
\begin{aligned}
\mathbf{EG} \ = \ & \sum_{n=0}^{\infty} \mathrm{P}(\mathrm{M}=\mathrm{n}) \times \\
& \left\{ h\left(n\right) \bar{S} + \left[\phi_0 - h\left(n\right)\right]\left[1 - \Psi\left(\phi_0 - h\left(n\right)\right)\right] \mathbf{E}\left(S_T \middle| M = n\right) \right\} \\
= \ & \left[1 - \int_0^T f_{0,1}\left(t\right) \mathrm{d}t \right] \phi_0 \cdot \left[1 - \Psi(\phi_0)\right] \mathbf{E}\left(S_T \middle| M = 0\right) \\
& + \sum_{n=1}^{n^*-1} \left[\int_0^T f_{0,n}(t)\mathrm{d}t - \int_0^T f_{0,n+1}(t)\,\mathrm{d}t \right] \times \\
& \left\{ h\left(n\right)\bar{S} + \left[\phi_0 - h\left(n\right)\right]\left\{1 - \Psi\left[\phi_0 - h\left(n\right)\right]\right\}\mathbf{E}\left(S_T\middle| M = n\right) \right\} \\
& + \left[\int_0^T f_{0,n^*}\left(t\right)\mathrm{d}t \right] \phi_0\, \bar{S},
\end{aligned}
$$

and $\mathbf{E}\left(S_T \middle| M = 0\right)$ *and* $\mathbf{E}\left(S_T \middle| M = n\right)$ *for* $n \geq 1$ *are given by*

$$
\begin{aligned}
& \mathbf{E}\left(S_T \middle| M = 0\right) \\
= \ & \int_{-\infty}^{\ln(\bar{S}/S_0)} \exp(s)\, S_0\, g\left(s \middle| \max_{0 \leq u \leq T} [\ln(S_u/S_0)] < \ln(\bar{S}/S_0),\ n = 0 \right) \mathrm{d}s
\end{aligned}
$$

$$
\begin{aligned}
& \mathbf{E}\left(S_T \middle| M = n\right) \\
= \ & \int_0^T \frac{f_{0,n}(t)\left(1 - \int_0^{T-t} f_{n,n+1}(s)\mathrm{d}s\right)}{\int_0^T f_{0,n}(u)\,\mathrm{d}u - \int_0^T f_{0,n+1}(u)\,\mathrm{d}u} \times \\
& \left[\int_{-\infty}^{\ln(\bar{S}/S_{t_n})} \exp(s)\, S_t\, g\left(s \middle| \max_{t \leq u \leq T} [\ln(S_u/S_t)] < \ln(\bar{S}/S_t),\ n \right) \mathrm{d}s \right] \mathrm{d}t.
\end{aligned}
$$

3.5   Modeling of the Drift Component

Up to now the drift of the best bid price has been assumed to be a constant. This section completes the theoretical framework by modeling explicitly the impact of the peak size on the drift following the intuition that the disclosure of large order volumes has an adverse market impact. For this purpose we will model the drift μ_t as a function of the order imbalance B_t.

Similar to Brown (1997) we define the imbalance B_t of the order book as the number of shares displayed on the bid side divided by the sum of shares displayed on the bid side and the ask side. The imbalance coefficient is bounded by 1 (if no orders are stored on the ask side of the book) and by 0 (if the bid side is empty). The parameter is 0.5 if the ask volume equals the bid volume. Whenever a new peak shows up in the order book the displayed ask volume increases, which in turn reduces B_t.

To keep the setup tractable for exposition, we assume the following simplified scenario: The best bid price exhibits a zero drift $\mu_t \equiv \bar{\mu} = 0$ prior to the submission of the iceberg order $(t < t_0)$. Furthermore, suppose that $B_t \equiv \bar{B}$ for $t < t_0$.

As soon as the iceberg order is submitted to the market the symmetry of the order book starts varying. Suppose that each variation in the displayed volume of the iceberg order ϕ_{dp} influences the order book symmetry. The displayed volume of other orders remains constant over time. Thus, we are able to model the imbalance as a function of the displayed volume of the iceberg order only:

$$B_t\left(\phi_{dp}\right) = \frac{c}{d + \phi_{dp}}, \qquad (3.8)$$

where the parameters c (and d) denote the number of shares displayed on the bid side (on the bid side *and* the ask side) before the submission of the iceberg order.

The displayed volume ϕ_{dp} is equal to ϕ_p whenever a new peak is submitted to the order book. When the peak of the iceberg order receives a complete or partial fill the parameter ϕ_{dp} will be reduced. In our setup the displayed volume ϕ_{dp} depends on the number of times the limit was already hit. It can be calculated as

$$\phi_{dp}(n)$$

$$= \begin{cases} \min\left[\phi_0 - h(n), \right. \\ \left. \phi_p - \left\{\max\left[\left(\frac{n\phi_s}{\phi_a+\phi_p} - \left\lfloor\frac{n\phi_s}{\phi_a+\phi_p}\right\rfloor\right)(\phi_a + \phi_p) - \phi_a, 0\right]\right\}\right] & \text{if } n < n^* \\ \\ 0 & \text{else.} \end{cases}$$

If the displayed volume of submitted orders has some information content to the market, one would expect a positive relationship between past levels of B_t and future returns. We define μ_n by the recursion

$$\mu_{n+1} = \mu_n + \beta \cdot \left(B_{t_{n+1}} - B_{t_n}\right) \text{ for } n \geq 0$$

with the initial value

$$\mu_0 = \bar{\mu} + \beta \cdot \left(B_{t_0} - \bar{B}\right),$$

such that

$$\Delta\mu_n = \beta\Delta B_{t_n}. \tag{3.9}$$

Figure 3.5 illustrates this idea. Before time t_0 the drift μ_t is equal to the long-term mean $\bar{\mu}$. At time t_0 the first peak of the iceberg order appears in the book and the drift μ_t is reduced. At times t_3, t_4, t_5, and t_6 the first peak of the iceberg order receives partial fills, which goes along with small upward jumps in the drift μ_t. At time t_7 the first peak becomes completely filled and the second peak appears in the order book, which again causes a downward jump in the drift μ_t. At time t_n^* the iceberg order is completely filled and the drift μ_t reverts towards its long-term mean $\bar{\mu}$.

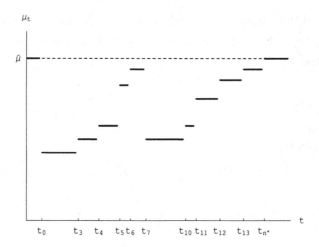

Fig. 3.5: Example for the alternating drift component in our model.

Note that μ_n is a deterministic function of the random variable n in our framework. Thus, we can rewrite equation (3.4) for the case where the drift depends on the displayed peak size:

$$f_{0,1}(t) = \frac{\ln\left(\bar{S}/S_0\right)}{\sigma\sqrt{2\pi t^3}} \times$$

$$\exp\left\{\frac{-\left[\ln\left(\bar{S}/S_0\right) - \left(\mu_0 - \sigma^2/2\right)t\right]^2}{2\sigma^2 t}\right\}. \qquad (3.10)$$

For the subsequent hitting times after the restart we get

$$f_{n-1,n}(t) = \frac{-\ln\left(1-\varepsilon\right)}{\sigma\sqrt{2\pi t^3}} \times$$

$$\exp\left\{\frac{-\left[-\ln\left(1-\varepsilon\right) - \left(\mu_{n-1} - \sigma^2/2\right)t\right]^2}{2\sigma^2 t}\right\}. \qquad (3.11)$$

This completes the introduction of the theoretical framework. We now turn to the numerical implementation of the open approach.

3.6 Numerical Results

To exemplify the formal analysis of the previous section the *open approach* is implemented using the MAN dataset for 61 trading days. The results are presented in the following.

3.6.1 Parameter Specification

Table 3.3: Estimated parameters for the MAN dataset for 61 trading days.

		No. of obs.	Mean	Std. dev.
ϕ_s	Average volume of all transactions	40,888	868.2	1,257.4
ϕ_a	Average displayed volume of all best ask quotes	158,607	1,554.4	1,752.0
ε	Average relative price difference between the best and the second best bid price	57,290	0.00090724	0.00105471
c	Average number of shares displayed on the bid side of the book	36,661	166,465.87	57,495.87
d	Average number of shares displayed on the bid side and on the ask side of the book	36,661	291,657.18	84,722.34

To implement the model a number of parameters need to be calibrated with order book data. Table 3.3 summarizes the results for our clinical sample. To estimate the parameters ϕ_s, ϕ_a, and ε we consider all observed transactions, and best and second best ask quotes with equal weights. For the calibration of the parameters c and d we use the order book data collected at intervals of 1 minute from 9:30 a.m. to 7:30 p.m. To estimate the parameter β we regress 60-minutes-ahead returns on changes of the order imbalance during the past 60 minutes, minute by minute. For this purpose we use best bid quotes and order book data collected at intervals of 1 minute from 9:30 a.m. to 7:30 p.m. We do not consider overnight returns for our analysis.

Table 3.4 reports the results.

Table 3.4: Estimated regression coefficient for equation (3.9).

		No. of obs.	Estimate	t-statistic
β	60-minute forecast intervals	29,341	116.647882	22.45

For the calibration of the volatility parameter we use best bid quotes collected at intervals of 15 minutes from 9:00 a.m. to 8:00 p.m. and do not consider overnight price changes. The estimation for the volatility parameter yields $\sigma = 0.7$. Furthermore, we set $S_0 = €\,28.55$, which is the closing price at March 28, 2002, the last day of our sample period.

3.6.2 Numerical Implementation

The computation of $f_{0,n}(t)$ requires the calculation of an n-th iterated convolution given by equation (3.6). In order to obtain $f_{0,n}(t)$ one needs to calculate $(n-1)$-dimensional integrals. To the best of our knowledge, closed form expressions are not available. Thus, we apply numerical approximations to these integrals. Employing conventional quadrature algorithms or Monte Carlo methods to compute high-dimensional integrals is very time consuming and thus not suitable for the dimensions under consideration in our framework. Therefore, we use interpolating cubic splines $s_{0,n}$ for $n \geq 2$ to approximate the convolutions in the following way:

$$f_{0,n}(t) \approx s_{0,n}(t),$$

such that

$$s_{0,n}(\hat{t}_k) = \int_0^{\hat{t}_k} s_{0,n-1}(\hat{t}_k - u) f_{n-1,n}(u) \, du \approx f_{0,n}(\hat{t}_k),$$

where $\hat{t}_k$ denotes the equally spaced spline knots.

Alternatively, one can also invert the Laplace transform of the density function.[6] The Laplace transform for the (defective) density function of the first hitting time τ_1 is well known (see, e.g., Karlin and Taylor, p. 362), and is given by

$$\mathbb{E} \exp(-\lambda \tau_1) = \exp \left\{ -\frac{\ln(\bar{S}/S_0)}{\sigma^2} \times \left[\sqrt{\left(\mu_0 - \frac{\sigma^2}{2}\right)^2 + 2\sigma^2 \lambda} - \left(\mu_0 - \frac{\sigma^2}{2}\right) \right] \right\}.$$

As the following sequences of hitting times τ_i for $i \geq 2$ are identically distributed their Laplace transform is given by

$$\mathbb{E} \exp(-\lambda \tau_i) = \exp \left\{ -\frac{\ln[\bar{S}/\bar{S}(1-\varepsilon)]}{\sigma^2} \times \left[\sqrt{\left(\mu_{i-1} - \frac{\sigma^2}{2}\right)^2 + 2\sigma^2 \lambda} - \left(\mu_{i-1} - \frac{\sigma^2}{2}\right) \right] \right\}.$$

The Laplace transform of the sum of the independent hitting times $t_n = \sum_{i=1}^n \tau_i$ is equal to the product of the corresponding exponential functions:

$$\mathbb{E} \exp(-\lambda t_n) = \exp \left\{ -\frac{\ln(\bar{S}/S_0)}{\sigma^2} \times \right.$$

---

[6]Numerical routines that do this job pretty fast are, for example, imsl_d_inverse_laplace from the IMSL C-Library or C06LAF/C06LBF from the Nag Fortran-Library.

$$\left[\sqrt{\left(\mu_0 - \sigma^2/2\right)^2 + 2\sigma^2\lambda} - \left(\mu_0 - \sigma^2/2\right) \right]$$
$$+ \sum_{i=2}^{n} - \left(\frac{-\ln\left(1-\varepsilon\right)}{\sigma^2} \right) \times$$
$$\left[\sqrt{\left(\mu_{i-1} - \sigma^2/2\right)^2 + 2\sigma^2\lambda} - \left(\mu_{i-1} - \sigma^2/2\right) \right] \Big\}.$$

3.6.3 Numerical Examples

For the first example, suppose that the investor wants to liquidate 10,000 MAN shares (approximately 1–2% of daily turnover) within 10 hours. Assume, furthermore, that ϕ_p has to be a multiple of 1,000 shares.

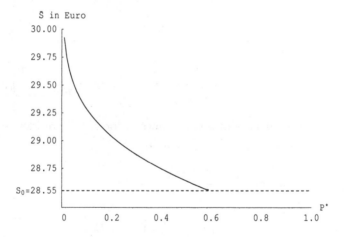

Fig. 3.6: Optimal limit $\bar{S}$ as a function of the probability P* that the iceberg order receives a complete fill. Other parameters: $T = 10$ hours, $\phi_0 = 10,000$ MAN shares.

Figure 3.6 represents the optimal limit $\bar{S}$ as a function of the probability P* that the iceberg order receives a complete fill. For P* $\leq$ 59%, the optimal limit is a monotonic decreasing function of P*. The optimal peak size remains at a constant level of 8,000 shares and is thus insensitive to changes of P*. Smaller or higher peak sizes reduce the value of the objective function, for

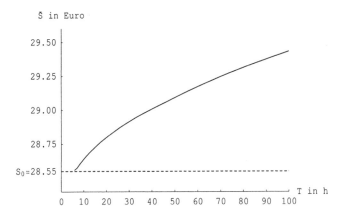

Fig. 3.7: Optimal limit $\bar{S}$ as a function of the final time horizon T. Other parameters: $P^* = 50\%$, $\phi_0 = 10{,}000$ MAN shares.

example by approx. 1% if $P^* = 30\%$ and $\phi_p = 1{,}000$ shares. If the limit is set
to € 28.56, i.e. the smallest possible value in this example, the probability
to observe a complete execution is still less than 60%.

The optimal peak size is significantly higher than peak sizes that were observed empirically in Section 3.3. Two reasons may explain the difference.
First, one may argue that the model systematically underestimates the negative price impact of displaying a large order volume in the book. There are
good reasons to believe that a variation of the order imbalance within entries close to the best quotes has a stronger impact on future returns than
changes of the order imbalance caused by an entry of an order that possesses
a more unfavorable price priority than the majority of other orders already
stored in the book. A redefinition of the order imbalance by weighting order book entries differently, depending on their price priority, might solve
this problem. Second, one may argue that market participants overestimate
the informational impact of revealing large orders in an open book. The
empirical exploration of these issues is left for further research.

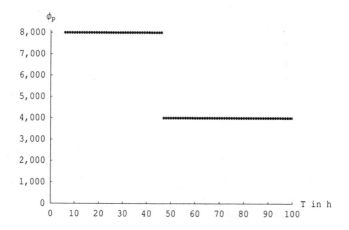

Fig. 3.8: Optimal peak size ϕ_p as a function of the final time horizon T. Other parameters: P* = 50%, ϕ_0 = 10,000 MAN shares.

In the next example we investigate the relationship between the final time horizon T and the optimal combination of limit and peak size. We set P* = 50% and ϕ_0 = 10,000 shares. Figures 3.7 and 3.8 report the results. If $T < 6$ hours, then the probability of receiving a complete fill is less than 50%, no matter which limit is assigned to the order. If $T \geq 6$ hours we can observe two beneficial effects for the originator of the iceberg order. First, as the final time horizon increases, the optimal order limit increases as well. Second, a longer time horizon allows for a reduction of the peak size. However, ϕ_p is not strictly monotonic decreasing in T. Instead we observe a step function. For $T \leq 46$ hours a peak size of 8,000 shares is optimal, for $T > 46$ the optimal peak size is 4,000 shares.

In the last example (see Figures 3.9 and 3.10) we analyze the relationship between the initial position ϕ_0 and the optimal pairs of ϕ_p and $\bar{S}$. We set P* = 25% and T = 100 hours. Figure 3.9 corroborates the hypothesis that if more shares have to be liquidated within the same period of time the limit has to be lowered to keep the execution probability at the same level. Furthermore, an increase in ϕ_0 tends to result in higher peak sizes,

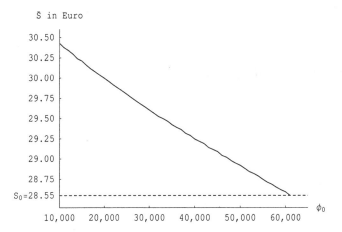

Fig. 3.9: Optimal limit $\bar{S}$ as a function of ϕ_0. Other parameters: $P^* = 25\%$, $T = 100$ hours.

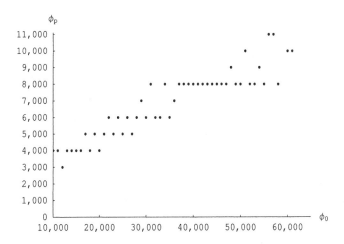

Fig. 3.10: Optimal peak size ϕ_p as a function of ϕ_0. Other parameters: $P^* = 25\%$, $T = 100$ hours.

as we can observe in Figure 3.10. However, in some cases the optimal peak size decreases if the initial position is raised. At the first moment this may seem somehow counterintuitive. The main reason for this phenomenon can be found in the discrete setup of the order execution process. Whenever the limit $\bar{S}$ is hit, a fixed transaction size ϕ_s is processed. At the n^*-th hit the last part of the iceberg order, which is given by $\phi_0 - h(n^* - 1)$, becomes executable. However, if $\phi_0 - h(n^*-1) \ll \phi_s$ a small reduction in ϕ_p would not change n^* but would increase the drift component μ_t and thus the probability that $t_n < T$.

3.7   Summary and Conclusion

This chapter introduces a setup that allows the determination of the optimal combination of limit and peak size of an iceberg order, given a large position in a security that should be liquidated within a finite time horizon. The framework balances the direct advantage of a large peak size that leads to a better time priority of an iceberg order and the adverse informational impact of revealing large order volumes in an open order book. Furthermore, it assesses the tradeoff between the order limit and the execution probability of the iceberg order. We have presented two approaches to incorporate the execution risk of an iceberg order. The so-called *self-contained approach* assumes that the unexecuted part is liquidated by a market order. The *open approach* is far more flexible as it does not require any assumption concerning the liquidation of the unexecuted part. It identifies the optimal combination of limit and peak size, given a minimum probability of complete order execution. Using real-world order book data we illustrate how the *open approach* can be implemented and explore major properties of the model by modifying input parameters.

To our knowledge, this framework is the first analytical approach that investigates the tradeoff between limit and peak size of an iceberg order, on the one hand, and the resulting execution probability, on the other. This chapter is written in search of a stylized model that is able to illustrate the interaction between observable market variables and order specific parameters that are important to analyze iceberg orders as a trading instrument.

The modeling of the best bid price by a Brownian motion or the assumption of constant parameters for order imbalance, transaction size, order flow and the price difference between the best and the second best price are, of course, approximations as the standard deviations in Table 3.3 clearly indicate. These simplifications allow us to keep the number of stochastic variables to the minimum required to illustrate the discussed trade-off in a simple way. Further research may focus on introducing more freedom from determinism by modeling more sources of risk, for example, in a simulation-based approach and comparing the empirical performance of the different models. Furthermore, although certainly challenging from a technical point of view, the investigation of dynamic approaches seems highly relevant from an empirical perspective, since many market participants pursue dynamic instead of static limit-setting strategies as shown in Table 3.1 in Section 3.3.

3.8 Appendix: Conditional Density g of $\ln\left(S_T/S_{t_n}\right)$

The conditional density g of $\ln\left(S_T/S_{t_n}\right)$ is given by

$$g\left(s\mid \max_{t_n \leq u \leq T} \ln(S_u/S_{t_n}) \; < \; \ln(\bar{S}/S_{t_n}), n\right)$$

$$= \frac{\psi_t\left(s, \ln\left(\bar{S}/S_{t_n}\right)\right)}{P\left(\ln\left(\max_{t_n \leq u \leq T} S_u \Big/ S_{t_n}\right) < \ln\left(\bar{S}/S_{t_n}\right)\right)},$$

where

$$\psi_t\left(x, y\right) = (1/\sigma)\exp\left[\left(\mu^* - \sigma^2/2\right)x/\sigma^2\right.$$
$$\left. - \left(\mu^* - \sigma^2/2\right)^2 (T - t_n)/2\sigma^2\right] \delta\left(x/\sigma, y/\sigma\right),$$

$$\delta\left(x, y\right) = \left[\varphi\left(x\left(T - t_n\right)^{-1/2}\right)\right.$$
$$\left. - \varphi\left(\left(x - 2y\right)\left(T - t_n\right)^{-1/2}\right)\right]\left(T - t_n\right)^{-1/2},$$

and

$$P\left(\ln\left(\max_{t_n \leq u \leq T} S_u \Big/ S_{t_n}\right) \; < \; \ln\left(\bar{S}/S_{t_n}\right)\right)$$

$$= \Phi\left(\frac{\ln\left(\bar{S}/S_{t_n}\right) - (\mu^* - \sigma^2/2)(T - t_n)}{\sigma\sqrt{(T - t_n)}}\right)$$
$$- \exp\left[2\left(\mu^* - \sigma^2/2\right)\ln\left(\bar{S}/S_{t_n}\right)/\sigma^2\right] \times$$
$$\Phi\left(\frac{-\ln\left(\bar{S}/S_{t_n}\right) - (\mu^* - \sigma^2/2)(T - t_n)}{\sigma\sqrt{(T - t_n)}}\right),$$

where $\varphi(z)$ denotes the standard normal density function and $\Phi(z)$ the standard normal cumulative distribution function. If the drift is a constant, as assumed in Section 3.4, set $\mu^* = \mu$. If the drift is modeled as a time-dependent variable, as proposed in Section 3.5, replace μ^* by μ_{t_n}. For the derivation of the respective formulas, see, e.g., Harrison (1990), pp. 1–16.

References

Aitken, M. J., P. Brown, and T. Walter (1996): Infrequent trading and firm size as explanation for the intra-day patterns in returns on SEATS, Working Paper, Securities Industry Research Centre of Asia-Pacific.

Aitken, M. J., H. Berkman, and D. Mak (2001): The use of undisclosed limit orders on the Australian Stock Exchange, *Journal of Banking & Finance* 25, 1589–603.

Almgren, R. and N. Chriss (2000): Optimal execution of portfolio transactions, *Journal of Risk* 3 (Winter 2000/2001), 5–39.

Bertsimas, D. and A. W. Lo (1998): Optimal control of execution, *Journal of Financial Markets* 1, 1–50.

Brown, P., D. Walsh, and A. Yuen (1997): The interaction between order imbalance and stock price, *Pacific-Basin Finance Journal* 5, 539–57.

Cho, J. W. and E. Nelling (2000): The probability of limit-order execution, *Association for Investment Management and Research*, September/October, 28–33.

Chordia, T. and A. Subrahmanyam (2002): Order imbalance and individual stock returns, Working Paper, Emory University, Atlanta, USA.

Copeland, T. E. and D. Galai (1983): Information effects on the bid–ask spread, *Journal of Finance* 38, 1457–69.

D'Hondt, C., R. De Winne, and A. François-Heude (2003): Hidden orders on Euronext: Nothing is quite as it seems . . . , Working Paper, FUCaM-Catholic University of Mons, Belgium.

Dubil, R. (2002): Optimal liquidation of large security holdings in thin markets, Working Paper, University of Connecticut, Storrs, USA.

Harrison, M. J. (1990): *Brownian Motion and Stochastic Flow Systems*, Krieger Publishing Company, Malabar, Florida, USA.

Hisata, Y. and Y. Yamai (2000): Research toward the practical application of liquidity risk evaluation methods, *Monetary and Economic Studies*, December, 83–128.

Karlin, S. and H. M. Taylor (1975): *A First Course in Stochastic Processes*, Academic Press, San Diego.

Lo, A. W., A. C. MacKinlay, and J. Zhang (2002): Econometric models of limit-order execution, *Journal of Financial Economics* 65, 31–71.

Mönch, B. (2003): Optimal liquidation strategies, Working Paper, Goethe University, Frankfurt am Main, Germany.

Pardo, Á. and R. Pascual (2003): On the hidden side of liquidity, Working Paper, Universidad de las Islas Baleares, Palma de Malorca, Spain.

Parlour, C. A. (1998): Price dynamics in limit order markets, *Review of Financial Studies* 11, 789–816.

Ranaldo, A. (2004): Order aggressiveness in limit order markets, *Journal of Financial Markets* 7, 53–74.

Wald, J. K. and H. T. Horrigan (2001): Optimal limit order choice, Working Paper, forthcoming 2006 in *Journal of Business* 79.

Zhang, P. G. (1998): *Exotic Options*, World Scientific Publishing, Singapore.

Lecture Notes in Economics and Mathematical Systems

For information about Vols. 1–459
please contact your bookseller or Springer-Verlag